Project Mercury

A Chronology

NASA SP-4001

**Prepared by James M. Grimwood,
Historical Branch, Manned Spacecraft Center, Houston, Texas,
as MSC Publication HR-1**

Office of Scientific and Technical Information

NATIONAL AERONAUTICS AND SPACE ADMINISTRATION

Washington, D.C. 1963

Updated February 13, 2006
Steven J. Dick, NASA Chief Historian
Steve Garber, NASA History Web Curator

Project Mercury

A Chronology

Published as NASA Special Publication-4001

Prepared by

James M. Grimwood

Table of Contents

Foreword

Preface

Acknowledgments

List of Illustrations

Introduction

Part I: Major Events leading to Project Mercury

- Part I (A) March 1944 through December 1957
- Part I (B) January 1958 through October 1, 1958

Part II: Research and Development Phase of Project Mercury

- Part II (A) October 3, 1958 through December 1959
- Part II (B) January 1960 through May 5, 1961

Part III: Operational Phase of Project Mercury

- Part III (A) May 5, 1961 through May 1962
- Part III (B) June 1962 through June 12, 1963

Appendices

- **Appendix 1: Project Mercury History**

   Appendix 2: Project Mercury Test Objectives

   Appendix 3: Project Mercury Flight Data Summary

   Appendix 4: Launch Site Summary, Cape Canaveral and Wallops Island

   Appendix 5: Project Mercury Budget Summary

   Appendix 6: Location of Mercury Spacecraft and Exhibit Schedule

Appendix 7: Launch Vehicle Deliveries to Cape Canaveral

Appendix 8: Key Management Progression Involving Project Mercury

Appendix 9: Contractors and Subcontractors Supporting Project Mercury

Appendix 10: Government Agencies Supporting Project Mercury

Foreword

Project Mercury is now history. In its short span of four years, eight months, and one week as the Nation's first manned space flight program, Mercury earned a unique place in the annals of science and technology. The culmination of decades of investigation and application of aerodynamics, rocket propulsion, celestial mechanics, aerospace medicine, and electronics, Project Mercury took man beyond the atmosphere into space orbit. It confirmed the potential for man's mobility in his universe. It remains for Projects Gemini and Apollo to demonstrate that potential.

Project Mercury was not only a step in the history of flight technology, it was a major step in national commitment to space research and exploration and to man's struggle to fly. One has only to contrast it with the Wright Brothers' achievements of sixty years ago, when two meticulous men, with a bicycle shop, a handmade wind tunnel, determination and industriousness, and little financial means or support, accomplished controlled, powered flight. The austere contrast of the Wrights or of Professor Goddard's rocket work with today's Government-sponsored, highly complex space program, involving thousands of persons and hundreds of Federal, industrial, and university activities, is eloquent testimony to the new prominence of science and technology in our daily lives. The evolution and achievements of Project Mercury offer an outstanding example of a truly national effort in the advancement of knowledge and its application.

The Project Mercury story must be examined in the full context of its fundamental features - scientific, engineering, managerial - in the dynamic human environment of national and international life. Indeed, the national commitment to Project Mercury and its successors requires a valid perspective on the potential accomplishments of science and technology as well as on the response of a democratic society to the challenges of its day.

This chronology of Project Mercury represents only a beginning on the full history, just as Mercury was only a first step in the development of American space transportation. No chronology is a history. This volume is but a preface to what is yet to come. Yet it offers us a catalog of processes by which man progresses from ideas originating in the human mind to the physical devices for man's travel to the moon and beyond.

Hugh L. Dryden

Deputy Administrator

Preface

Project Mercury stands as the free world's first program for manned exploration of space. History will show that it has been remarkably successful for a number of reasons. Primarily, all of the technical objectives necessary to the successful completion of the program were accomplished. Also, Mercury experience has provided this nation with the capability to implement and manage future projects on a level of quality and effectiveness that would otherwise have been impossible. Possibly of greater significance is the fact that Project Mercury was conceived and carried out solely for peaceful purposes, and all major events have been fully documented in the public news media, including television coverage of each manned launch from Cape Canaveral.

It is remarkable that the original goal of Mercury, that of orbiting a man in space and returning him safely to earth, was accomplished in just 3 years after the prime contract was awarded. This element of the program's success is especially significant when compared to development efforts for more conventional manned aircraft in which development and qualification periods of 5 or more years are not uncommon. The rapid pace with which the critical program milestones were completed was possible only through the dedicated efforts of many thousands of people. Because of the success in meeting prescribed technical objectives and the reliable operation of the spacecraft and launch vehicle systems it was possible to eliminate certain qualification flights early in the program and broaden the original scope of Mercury into the recent and final manned 1-day mission of 22 orbits or 34 hours duration. The valuable experience gained in the design, development, and operation of the Mercury spacecraft, as well as in management of such a program, has already resulted in a profound effect on the Gemini and Apollo projects and will continue to do so to an even greater extent.

This document presents a brief but accurate chronology of important events throughout the Mercury program and attests to the rapid pace at which the Mercury development and operation were carried out. Many of the critical decisions which were later significantly to affect the direction of the program are mentioned, and the manned flights, from the first sub-orbital mission of May 5, 1961, to the final orbital mission conducted on May 15 and 16, 1963, are documented. Project Mercury is now history, and only time will allow a complete assessment of its full impact on this nation's technology and contribution in expanding the space frontier. But it can be stated without reservation that this project will be remembered as one of the outstanding technical achievements that this country has contributed to world history.

Kenneth S. Kleinknecht
Manager, Mercury Project

Acknowledgments

Material for this document was accumulated from widespread sources, and for this reason the author is indebted to a number of individuals. In fact, several persons contributed to such a degree that formal recognition is warranted. These were as follows: Mr. Paul E. Purser, Special Assistant to Director, MSC; Mrs. Phoncille De Vore of Mr. Purser's office; Lt. Colonel John A. Powers, Public Affairs Officer, MSC; Mr. William M. Bland, Deputy Manager of Project Mercury, MSC; Mr. Robert W. Fricke, Technical Information Division, MSC; Miss Retha Shirkey, Librarian, MSC; Mr. Ralph Shankle, Public Affairs Office, MSC; Mr. David S. Akens, Marshall Space Flight Center; Mr. Alfred Rosenthal, Goddard Space Flight Center; Dr. Eugene M. Emme, the NASA Historian; and Mrs. Helen Wells of the NASA Headquarters Historical Office. Especial thanks is given to Mrs. Frankie J. Fisher, the branch secretary, for typing, layout, and research.

JMG

List of Illustrations

Frontispiece: Mercury spacecraft with escape tower

Figure 1: Closeup view of Recruit Escape Rocket and full-scale spacecraft

Figure 2: Mercury spacecraft in orbit: Artist's conception

Figure 3: Little Joe on launcher at Wallops during checkout

Figure 4: Pilot egress trainer

Figure 5: Manufacture of Mercury spacecraft at McDonnell plant, St. Louis, Mo.

Figure 6: Shadowgraph of spacecraft model in Ames Supersonic Free-Flight Pressurized Range

Figure 7: Equipment installation in the parachute canister

Figure 8: McDonnell mockup of Mercury spacecraft including Atlas adapter and escape system

Figure 9: Escape rocket motor

Figure 10: The seven Mercury astronauts

Figure 11: Scale model of escape tower configuration

Figure 12: Honeycomb structure partially to absorb impact force

Figure 13: Spacecraft and escape system configuration

Figure 14: Human centrifuge used in Mercury astronaut training program

Figure 15: Spacecraft interior arrangement

Figure 16: Astronaut survival equipment stowed in Mercury spacecraft

Figure 17: Recovery test spacecraft showing recovery aids

Figure 18: Main 63-foot ringsail parachute

Figure 19: White room in Hanger S at Cape Canaveral

Figure 20: Spacecraft with McDonnell designed escape system ready for firing at Wallops Island

Figure 21: Spacecraft reaction control system

Figure 22: Vehicle for drogue parachute test at NASA Flight Research Center

Figure 23: Flight plan for drogue parachute tests at NASA Flight Research Center

Figure 24: Big Joe on launch pad at Cape Canaveral for ballistic reentry flight test

Figure 25: Spacecraft instrument control panel

Figure 26: Three-axis hand controller

Figure 27: Mercury spacecraft heat shield after reentry

Figure 28: Mercury Control Center at Cape Canaveral

Figure 29: Plaster forms of contour couches

Figure 30: Reaction control thrust chamber

Figure 31: Rhesus monkey, "Miss Sam," being placed in container for LJ-1B flight

Figure 32: Manufacture of Mercury spacecraft at McDonnell plant, St. Louis, Mo.

Figure 33: Landing shock attenuation system

Figure 34: Astronauts in weightless flight in C-131 aircraft

Figure 35: Mercury altitude test chamber in Hanger S, Cape Canaveral

Figure 36: Posigrade rocket motor

Figure 37: Pressure suit worn by Alan Shepard on first manned suborbital space flight

Figure 38: Mercury environmental control system

Figure 39: Mercury-Atlas 1

Figure 40: Mobile pad egress tower (cherry picker)

Figure 41: Mercury spacecraft and astronaut Shepard being recovered by Marine Corps helicopter

Figure 42: Tower jettison rocket motor

Figure 43: Computers used in Mercury orbital track at Goddard Space Flight Center

Figure 44: Spacecraft antennas

Figure 45: Chimpanzee, "Ham," flown in Mercury-Redstone 2 suborbital flight

Figure 46: Impact attenuation

Figure 47: Atlas launch vehicle 100-D delivered to Cape Canaveral for Mercury-Atlas 3 flight

Figure 48: MA-8 orbital track: Mercury worldwide tracking network

Figure 49: Tracking site at Kano, Nigeria, Africa

Figure 50: Scout launch vehicle proposed to test Mercury worldwide tracking network

Figure 51: Mercury-Redstone 3: First manned suborbital space flight

Figure 52: Mercury-Redstone 3 flight profile

Figure 53: Freedom 7 returned by helicopter to *USS Lake Champlain*

Figure 54: Key personnel in Mercury Control Center at Cape Canaveral

Figure 55: Normal Mercury-Atlas orbital mission sequence

Figure 56: Auxiliary flotation collar

Figure 57: Production of Atlas launch vehicles at Convair Astronautics plant

Figure 58: Chimpanzee, "Enos," flown in Mercury-Atlas 5 two-orbit mission

Figure 59: Scuba divers prepare for recovery of Mercury spacecraft

Figure 60: Mercury-Atlas 6: First manned (Glenn) orbital flight

Figure 61: Balloon experiment

Figure 62: Zero-gravity experiment

Figure 63: Astronaut couch modifications

Figure 64: MA-8 ditty bag contents

Figure 65: Astronaut departs transfer van for Mercury-Atlas gantry

Figure 66: Atlas launch vehicle 130-D (MA-9) undergoing inspection at Cape Canaveral

Figure 67: Flight pressure suit of astronaut L. Gordon Cooper used in MA-9, 22-obit mission

Figure 68: Astronaut L. Gordon Cooper prepares for insertion in Faith 7 (MA-9)

Introduction

A decision by National Aeronautics and Space Administration Headquarters in October 1961 extended the Mercury program by adding 1-day missions after three- and six-orbit flights. Also, during the same year, follow-on manned space programs, later known as Projects Gemini and Apollo, began to take form. These events were rather unusual, for here was program expansion on a higher level of difficulty prior to the time that the basic objectives of Project Mercury, the launch and safe return of a man from earth orbit, had been attained. Obviously, Project Mercury, first guided by the Space Task Group and then by the Manned Spacecraft Center (the successor organization), had built up a high confidence factor as to the potential success of the space venture. To a large degee, this action was graphically supported at that point in time by the highly successful suborbital flights of Alan Shepard and Virgil Grissom and the orbital flight of the "mechanical astronaut."

Project Mercury's formal program approval date was October 7, 1958, and 3 years and 2 weeks from the award of the development and production contract by NASA to the McDonnell Aircraft Corporation of St. Louis, Missouri, the orbital flight of John Glenn aboard the Mercury spacecraft, "Friendship 7," transpired. When this uncommonly brief time scale is compared with other major programs of national note and urgency, the question of how man was committed so soon to orbital flight is certain to be posed.

The key to this phenomenal success was concurrency of effort. That is, all facets of the program leading to manned space flight were guided along a simultaneous route and not by the concept of qualifying each phase before development work began on another. From the outset, work was being accomplished on all components of the spacecraft, adapting the launch vehicles, readying the worldwide tracking network, selecting and training astronauts, and developing ground support equipment for systems checkout and astronaut training. No detail was too small to warrant the attention of scientists and engineers who were charged with making the awesome decisions that would commit man to orbital flight. Every organization that had acquired any technical proficiency or had built up a capability in a particular field that could be applied to the space program was visited, and arrangements were made for assistance, facilities, or the use of equipment. Also, the test and reliability program to which Mercury hardware was subjected was exhaustive and thorough. In fact, this unusually close attention refutes the "crash program" connotation often cited. The term "accelerated" more aptly describes the effort. That the managers were not swayed toward a crash program even in the face of an American public anxiously awaiting the advent of manned space flight, was unusual.

There were a number of catalysts which created the conditions leading to the approval of the Mercury project, and many of these circumstances and events contributed directly to the goal of attaining manned space flight. Shortly after World War II, experimental missile tests were conducted in the White Sands, N. Mex. area to altitudes beyond the sensible atmosphere. During this same period, rocket aircraft research was initiated with the objective of piercing the sound barrier. Then from the early to the mid-fifties the National Advisory Committee for Aeronautics and industry scientists and engineers made the assault on the thermal barrier to resolve the reentry problem for the ballistic missile. These excellent mediums of research formed a natural progression for the NACA to attack the problems of manned space flight. Another factor contributing to the growing interest in the national space program was the planning and research that was devoted to the artificial earth satellite program for the International Geophysical Year. Then the flight of Sputnik I in 1957 furnished the "yeast" necessary for the American public to support a manned space flight project. Finally, the Atlas launch vehicle had reached a point in development at which serious consideration could be given for its application to manned space flight. At that time the Atlas was the only American launch vehicle capable of lifting a payload for the manned orbital requirements.

This document chronicles the three major phases of the Mercury program - conception, research and development, and operation. Even in this brief form, the reader can readily observe the meticulous attention to detail that was given by personnel of the NASA, other Government agencies, and American industry associated with the conduct of the program to assure mission success in our first manned step in space.

James M. Grimwood

MSC Historian

PART I (A)
Major Events Leading to Project Mercury

March 1944 through December 1957

1944

March 16

At a National Advisory Committee for Aeronautics (NACA) seminar, in Washington, D.C., with Air Force and Navy personnel attending, NACA personnel proposed a jet-propelled transonic research airplane be developed. This proposal ultimately led to the "X" series research airplane projects.
> Eugene M. Emme, *Aeronautics and Astronautics: An American Chronology of Science and Technology in the Exploration of Space: 1915 - 1960* (Washington: NASA HHR-3, 1961), p. 47. Hereinafter cited as Emme, *Aeronautics and Astronautics, 1915 - 1960*.

December 9

A meeting was held at the Langley Aeronautical Laboratory, Langley Air Force Base, Virginia, to discuss the formation of an organization that would devote its efforts to the study of stability and maneuverability of high-speed weapons (guided missiles). From the outset, work was pointed toward supersonic flight testing. In early 1945, Congress was asked for a supplemental appropriation to fund the activation of such a unit, and in the spring of that year the Auxiliary Flight Research Station (AFRS - later known as the Pilotless Aircraft Research Division) was opened on Wallops Island, Virginia, with Robert R. Gilruth as its director. On July 4, 1945, the AFRS launched its first test vehicle, a small two-stage, solid-fuel rocket to check out the installation's instrumentation.
> Data supplied by Joseph A. Shortal, Chief, Applied Materials and Physics Division (formerly PARD), Langley Research Center, May 28, 1963.

1944 (during the year)

Congress appropriated funds to carry out a rocket aircraft research program. The National Advisory Committee for Aeronautics, the Air Force, and the Navy were designated participating members.
> Charles V. Eppley, *The Rocket Research Aircraft Program: 1946 - 1962* (Air Force Flight Test Center, Edwards Air Force Base, Calif.), p. 1. Hereafter cited as Eppley, *Rocket Research Aircraft Program: 1946 - 1962*.

1946

March

The Army Air Forces established Project RAND, which in part included the study of satellite applications.
    Emme, *Aeronautics and Astronautics: 1915 - 1960*, p. 53.

May 8

The Chief of Naval Operations directed the Navy's Bureau of Aeronautics to make preliminary investigations in the field of earth satellite vehicles.
    Emme, *Aeronautics and Astronautics: 1915 - 1960*, p. 54.

May 12

Project RAND filed a report entitled "Preliminary Design of an Experimental World Circling Space Ship," which indicated the technical feasibility of building and launching an artificial satellite.
    House Report No. 360, *Military Astronautics (Preliminary Report)*, 87th Congress, 1st Session, p. 2.

1947

October 14

The XS-1 rocket plane made the first supersonic manned flight by traveling 700 miles per hour (mach 1.06 at 43,000 feet altitude) over Muroc Dry Lake, California, with Captain Charles E. Yeager at the controls. The sound barrier was broken.
    Eppley, *Rocket Research Aircraft Program: 1946 - 1962*, p. 6.

October (during the month)

Due to the number of competing study contracts on satellites that were being submitted, the Department of Defense assigned responsibility to coordinate this work to the Committee on Guided Missiles of the Research and Development Board.
    House Rpt. 360, 87th Cong., 1st Sess., p. 2.

1948

January 15

General Hoyt S. Vandenberg, Vice Chief of Staff, United States Air Force, approved a policy calling for the development of earth satellites at the proper time.
>Emme, *Aeronautics and Astronautics: 1915 - 1960*, p. 59.

June 11

A V-2 designated "Albert" in honor of its passenger was launched at White Sands, New Mexico. Albert, the first American primate in space, died of suffocation. On June 6, 1949, Albert II was launched into space but died on impact. During 1949 two other flights of this type were conducted. In each case, the primate survived the flight, but succumbed before his capsule was located.
>David S. Akens, *Origins of Marshall Space Flight Center*, pp. 8-9. Hereinafter cited as Akens, *Origins of MSFC*.

July 13

Convair's MX-774 test vehicle, later designated the Atlas and used as a launch vehicle in the Mercury program, was test-fired for the first time.
>Emme, *Aeronautics and Astronautics: 1915 - 1960*, p. 60.

December 29

The first Secretary of Defense, James V. Forrestal, in his initial report to President Harry Truman, included a brief item indicating that the earth satellite program, which was being carried out independently by the military services, was assigned to the Committee on Guided Missiles for coordination.
>House Rpt. 360, 87th Cong., 1st Sess., p. 2.

1949

May 11

President Harry S. Truman signed a bill authorizing the missile test range, which is now the Atlantic Missile Range at Cape Canaveral, Florida.
>Emme, *Aeronautics and Astronautics: 1915 - 1960*, p. 62.

1951

January 16

The Government decided to resume MX-774 studies, and the project was then designated the Atlas. Several test vehicles had been fired in 1948 and 1949, after which the Convair MX-774 (Atlas) missile project had been shelved. The company, however, had continued to fund a research program.
    House Report No. 67, *A Chronology of Missile and Astronautic Events*, 87th Congress, 1st Session, p. 14.

September 20

The first successful recovery of animals from rocket flight in the Western Hemisphere was made when a monkey and 11 mice survived an Aerobee launch to an altitude of 236,000 feet.
    Emme, *Aeronautics and Astronautics: 1915 - 1960*, p. 68.

1952

January 30

An NACA report was issued covering several projects and proposals for the flight of manned and unmanned vehicles to altitudes above the earth where atmospheric density was very low. The substance of these reports was presented at the June 24, 1952, meeting of the Committee on Aerodynamics. After the presentation, committee member Robert J. Woods recommended that basic research be initiated on the problems of space flight and stated that the NACA was the logical organization to carry on this work. To accomplish this task, a small working group was established to analyze the available information on the subject of space flight. The objective of this group was to arrive at a concept of a suitable manned test vehicle that could be constructed within 2 years.
    Minutes of Meeting, NACA Committee on Aerodynamics, June 24, 1952.

May 16

The Special Committee for the International Polar Year (later designated the International Geophysical Year), was established.
    House Rpt. 67, 87th Cong., 1st Sess., p. 16.

June 18

H. Julian Allen of the NACA Ames Aeronautical Laboratory, Moffett Field, California, conceived of the blunt nose principle for reentry vehicles. On this date Allen stated he had determined that the blunt form would be suitable for any body reentering the earth's atmosphere. This principle was first used on the

intercontinental ballistic missile nose cone and was later incorporated into the configuration of the Mercury spacecraft.
  Emme, *Aeronautics and Astronautics: 1915 - 1960*, p. 69; Information supplied by Jack Talmadge, Ames Research Center, May 28, 1963.

June 24

The NACA Committee on Aerodynamics recommended that NACA increase its research efforts on the problem of manned and unmanned flight at altitudes between 12 and 50 miles and at speeds of mach 4 through 10. As a result of this recommendation, the Langley Aeronautical Laboratory began preliminary studies on this project and immediately identified several problem areas. Two of these areas were aerodynamic heating and the achievement of stability and control at very high altitudes and speeds. Of the two, Langley considered aerodynamic heating to be the more serious, and, until this problem was resolved, the design of practical spacecraft impractical. (See January 30, 1952, entry).
  Minutes of Meeting, NACA Committee on Aerodynamics, June 24, 1952.

June (during the month)

The Navy's Johnsville, Pennsylvania, human centrifuge began operations. This installation was later designated the Aviation Medical Acceleration Laboratory (AMAL) and was used extensively in the training of the Mercury astronauts.
  House Rpt. 67, 87th Cong., 1st Sess., p. 16.

1952 (during the year)

The NACA Langley Aeronautical Laboratory Pilotless Aircraft Research Division started the development of multistage, hypersonic-speed, solid-fuel, rocket vehicles. These vehicles were used primarily in aerodynamic heating tests at first and were then directed toward a reentry physics research program.
  Message, NASA Space Task Group to NASA Hq., July 5, 1960.

Between 1952 - 1956

Personnel of NACA Langley and Ames Aeronautical Laboratories were engaged in research on aerodynamic characteristics of reentry configurations. Knowledge acquired from these efforts along with those of industry and the military services was used in Project Mercury, proved the ablation theory for the Army's Jupiter missile development program, and was used in the Air Force intercontinental ballistic missile nose cone reentry program.
  Message, NASA Space Task Group to NASA Hq., July 5, 1960.

1953

July 30

Preliminary studies were completed by C. E. Brown, W. J. O'Sullivan, Jr., and C. H. Zimmerman at the Langley Aeronautical Laboratory relative to the study of the problems of manned space flight and a suggested test vehicle to investigate these problems. One of the possibilities considered from the outset of the effort in mid-1952 was modification of the X-2 airplane to attain greater speeds and altitudes of the order of 200,000 feet. It was believed that such a vehicle could not only resolve some of the aerodynamic heating problems, but also that the altitude objective would provide an environment with a minimum atmospheric density, representing many problems of outer space flight. However, there was already a feeling among many NACA scientists that the speed and altitude exploratory area should be raised. In fact, a resolution to this effect, presented as early as July 1952, stated that ". . . the NACA devote . . . effort to problems of unmanned and manned flights at altitudes from 50 miles to infinity and at speeds from mach 10 to the velocity of escape from the earth's gravity." The Executive Committee of NACA actually adopted this resolution as an objective on July 14, 1952.
>Letter, NACA to High Speed Flight Research Station, Subj: Discussion of Report on Problems of High Speed, High Altitude Flight, and Consideration of Possible Changes to the X-2 Airplane to Extend its Speed and Altitude Range, July 30, 1953.

August 20

The first Redstone missile was test-fired by the Army at Cape Canaveral, Florida. The Redstone, on which research and development had begun in 1950, was later used as a launch vehicle in the manned suborbital flights and in other development flights in Project Mercury.
>Emme, *Aeronautics and Astronautics: 1915 - 1960*, p. 72.

1954

May

The NACA determined the characteristics of what later became the X-15 rocket aircraft, one of the steps to manned space flight.
>Eppley, *Rocket Research Aircraft Program: 1946 - 1962*, p. 24.

June 25

In a meeting, Dr. Wernher von Braun, Frederick C. Durant III, Alexander Satin, David Young, Dr. Fred L. Whipple, Dr. S. Fred Singer, and Commander George W. Hoover agreed that a Redstone rocket with a Loki cluster as the second stage could launch a satellite into a 200-mile orbit without major new developments.

Project Orbiter was a later outgrowth of this proposal and resulted in the launching of Explorer I on January 31, 1958.
>House Rpt. 67, 87th Cong., 1st Sess., p. 19; Emme, *Aeronautics and Astronautics: 1915 - 1960*, p. 75; and James M. Grimwood, *History of the Jupiter Missile System* (Army Missile Command, Redstone Arsenal, Alabama, July 1962).

July 9

After 2 years' study of problems that might be encountered in manned space flight, a joint group - NACA, Air Force, and Navy - met in Washington to discuss the need for a hypersonic research vehicle and to decide on the type of aircraft that could attain these objectives. The NACA proposal was accepted in December 1954, and a formal memorandum of understanding was signed to initiate the X-15 project. Technical direction of the project was assigned to the NACA. On November 9, 1961, the X-15 reached its design speed of over 4,000 miles per hour and achieved partial space conditions on July 17, 1962, when it reached an altitude of 314,750 feet. By the latter date, the Mercury spacecraft had made two manned orbital flights.
>Eppley, *Rocket Research Aircraft Program: 1942 - 1946*, pp. 24, 44, 45.

August 7

The Air Force School of Aviation Medicine at Randolph Field, Texas, received the first specifically built space cabin simulator.
>House Rpt. 67, 87th Cong., 1st Sess., p. 19.

October 14

The first American four-stage rocket was launched by the Pilotless Aircraft Research Division of NACA's Langley Laboratory at Wallops Island.
>Emme, *Aeronautics and Astronautics: 1915 - 1960*, p. 76.

1955

March

Dr. Alan T. Waterman of the National Science Foundation presented President Dwight Eisenhower with a plan to implement the United States' portion of the International Geophysical Year satellite experiment.
>Emme, *Aeronautics and Astronautics: 1915 - 1960*, p. 79.

July 29

President Eisenhower endorsed the IGY proposal for the launching of small earth-circling satellites.
>Emme, *Aeronautics and Astronautics: 1915 - 1960*, p. 78.

The United States announced that it would launch earth satellites during the 18-month IGY (July 1957 through December 1958).
House Rpt. 67, 87th Cong., 1st Sess., p. 22.

September 9

Project Vanguard began operations. On this date the Department of Defense wrote a letter to the Department of Navy authorizing the Navy Research Laboratory to proceed with the Vanguard proposal. The objective of the program was to place a satellite in orbit during the IGY, and responsibility for carrying out the program was placed with the Office of Naval Research.
John P. Hagen, "The Viking and the Vanguard: History of Rocket Technology;" in special issue of *Technology and Culture* (Fall 1963).

The Department of Defense's Stewart Committee reviewed the alternatives for an IGY satellite program: wait for the development of an Atlas launcher, use a modified Redstone, or develop a rocket derived from the Viking missile. The committee voted seven to two in favor of abandoning Project Orbiter (Redstone) and developing Vanguard (the Viking derivative). Secretary Donald Quarles ruled with the committee majority in the Department of Defense Policy Committee, which approved the decision.

House Rpt. 67, 87th Cong., 1st Sess., p. 23; Grimwood, *History of the Jupiter Missile Program*; Akens, *Origins of MSFC*, pp. 38-40.

October 2

The National Academy of Sciences established a Technical Panel for Earth Satellite Program, with Richard E. Porter serving as chairman.
Emme, *Aeronautics and Astronautics: 1915 - 1960*, p. 79.

During 1955 - 1956

The NACA Langley and Ames Aeronautical Laboratories developed high-temperature jets, wind tunnels, and other facilities for use in materials and structures research at hypersonic speeds. These facilities provided, among other things, data proving that ablation was an efficient heat-protection method for reentry vehicles.
Message, NASA Space Task Group to NASA Hq., July 5, 1960.

1956

February 1

The Army Ballistic Missile Agency (ABMA) was activated at Redstone Arsenal, Huntsville, Alabama, to complete the development of the Redstone missile and to develop the Jupiter missile. The Redstone was later used in two Mercury manned suborbital flights, and in other research and development flights.
    Helen Joiner, *History of the Army Ballistic Missile Agency, 1 Feb.-30 June 1956.*

March

Project 7969, entitled "Manned Ballistic Rocket Research System," was initiated by the Air Force with a stated task of recovering a manned capsule from orbital conditions. By December of that year, proposal studies were received from two companies, and the Air Force eventually received some 11 proposals. The basis for the program was to start with small recoverable satellites and work up to larger versions. The Air Force Discoverer firings, which effected a successful recovery in January 1960, could be considered as the first phase of the proposed program. The Air Force program was based upon a requirement that forces no higher than 12g be imposed upon the occupant of the capsule. This concept required an additional stage on the basic or "bare" Atlas, and the Hustler, now known as the Agena, was contemplated. It was proposed that the spacecraft be designed to remain forward during all phases of the flight, requiring a gimballed seat for the pilot. Although the Air Force effort in manned orbital flight during the period 1956-58 was a study project without an approved program leading to the design of hardware, the effort contributed to manned space flight. Their sponsored studies on such items as the life-support system were used by companies submitting proposals for the Mercury spacecraft design and development program. Also, during the 2-year study, there was a considerable interchange of information between the NACA and the Air Force.
    House Rpt. 1228, *Project Mercury, First Interim Report*, 86th Congress, 2nd Session, p. 2; Comments by Clotaire Wood, NACA, Jan. 26, 1960, on Draft, NIS Meeting at ARDC Headquarters, June 19, 1958; Memo, Maxime A. Faget, NACA Langley, to Dr. Hugh Dryden, Director, NACA (no subject), June 5, 1958; Comments by Maxime A. Faget on "Outline of History of USAF Man-in-Space R&D Program," *Missiles and Rockets*, Vol. 10, No. 13 (Mar. 26, 1962), pp. 148-149.

May 3

The Air Force disclosed that a $41 million guided missile production facility would be built at Sorrento, California, for the Atlas launch vehicle. Convair was announced as the prime contractor.
    Emme, *Aeronautics and Astronautics: 1915 - 1960*, p. 82.

August 24

A five-stage, solid-fuel rocket test vehicle, the world's first, was launched to a speed of mach 15 by the NACA Langley Aeronautical Laboratory's Pilotless Aircraft Research Division.
    House Rpt. 67, 87th Cong., 1st Sess., p. 27.

October

NACA scientists were engaged in preliminary studies of the need for a follow-on, manned-rocket research vehicle to the X-15.
    Emme, Aeronautics and Astronautics: 1915 - 1960, p. 83.

November

Personnel of the Air Research and Development Command approached NACA officials on the possible cooperation of NACA in a research airplane project as a follow-on to the X-15 project. NACA agreed to consider the plan and directed its laboratories to initiate feasibility studies relative to the range of speed for the proposed vehicle and an estimate of the time frame in which the vehicle could be developed.
    NACA Study of the Feasibility of a Hypersonic Research Airplane, Sept. 3, 1957, p. 3.

1956 (during the year)

Personnel of the NACA were studying the possibilities of utilizing existing ballistic missile boosters, which were then under development, for manned orbital space flight.
    Letter, Paul E. Purser, MSC, to Mary Stone Ambrose, Policies and Regulation Branch, NASA Hq. (no subject), undated.

1957

January 14

The United States proposed before the United Nations Assembly that study be initiated toward international agreements assuring the use of outer space for peaceful purposes only.
    House Document No. 71, Message from the President of the United States, U.S. Aeronautics and Space Activities: January 1 to December 31, 1958, p. 18.

June 11

The first launch attempt of the Atlas was made at Cape Canaveral, Florida, but the missile exploded shortly after takeoff at an altitude of about 10,000 feet.
    George Alexander, "Atlas Accuracy Improves as Test Program is Completed," Aviation Week and Space Technology, Feb. 25, 1963, p. 54.

June 20

Two NACA groups focused their efforts on the problems involved in manned space flight. One group concerned themselves with performance of aircraft at high speeds and altitudes and with rocket research; the other group, with problems associated with hypersonic flight and reentry.
  Study, *NACA Research into Space*, Dec. 1957.

July (during the month)

A study was initiated by the Langley Aeronautical Laboratory on the use of solid-fuel upper stages to achieve a payload orbit with as simple a launch vehicle as possible. This was the beginning of the Scout test-vehicle concept.
  Emme, *Aeronautics and Astronautics: 1915 - 1960*, p. 87.

July - August

Alfred J. Eggers, Jr., of the NACA Ames Aeronautical Laboratory, worked out a semiballistic design for a manned reentry spacecraft.
  Emme, *Aeronautics and Astronautics: 1915 - 1960*, p. 87.

August 7

A Jupiter-C (test vehicle in the Jupiter missile development program), with a scale-model nose cone, was fired 1,200 miles down the Atlantic Missile Range. The nose cone, an ablative type, reached a peak altitude of over 600 miles, and its recovery was one of the proving steps of the ablative reentry principle. The nose cone was displayed by President Eisenhower to a nation-wide television audience on November 7, 1957.
  Army Capabilities in the Space Age, p. 26; Grimwood, *History of the Jupiter Missile System*; Emme, *Aeronautics and Astronautics: 1915 - 1960*, p. 87.

September 25

The second Atlas launch vehicle was destroyed in a launching attempt at Cape Canaveral, Florida.
  House Rpt. 67, 87th Cong., 1st Sess., p. 32.

October 4

The Union of Soviet Socialist Republics launched Sputnik I, the first artificial earth satellite. This event galvanized interest and action on the part of the American public to support an active role in space research, technology, and exploration.
  Emme, *Aeronautics and Astronautics: 1915 - 1960*, p. 91; Senate Hearings, 86th Congress, 2nd Session, Missiles, Space, and Other Major Defense Matters, Feb. 2-4, 8-9, March 16, 1960, p. 331. Also at this time, many leaders, Dr. Wernher von Braun, for example, made speeches on the "Impact of Sputnik" to American audiences anxious to

learn the meaning and to act to meet the requirements. For a concise statement on the subject see Appendix C, "The Public Impact of Early Satellite Launching" in Senate Rpt. 1014, *Project Mercury: Man-in-Space Program of the NASA*, p. 71.

October 14

The American Rocket Society presented President Eisenhower with a suggested program for outer space exploration. They proposed the establishment of an Astronautical Research and Development Agency similar to NACA and the Atomic Energy Commission. This agency would have responsibility for all space projects except those directly related to the military services. A list of proposed projects was presented at an estimated cost of $100 million per annum.
    House Rpt. 67, 87th Cong., 1st Sess., p. 33.

October 15-21

A "Round 3" conference involving studies for a follow-on to the X-15 program, which subsequently led to the X-20 Dyna Soar, was held at the Ames Aeronautical Laboratory. During the course of the meeting, Alfred J. Eggers, Jr., of Ames advanced several proposals for possible manned satellite vehicle development projects.
    Memo, Warren J. North to NASA Administrator, subject: Background of Project Mercury Schedules, Aug. 14, 1960.

November 8

Secretary of Defense Neil McElroy directed the Army to proceed with the launching of the Explorer earth satellites. This order, in effect, resumed the Orbiter project that had been eliminated from the IGY satellite planning program on September 9, 1955.
    Akens, *Origins of MSFC*, p. 45.

November 12-13

At a meeting of the NACA Subcommittee on Fluid Mechanics, it was stated that many aspects of space flight and astronautics would depend heavily on research advances in the field that had been broadly termed fluid mechanics. Research in this area involved internal and external gas flows associated with high-speed flights within the atmosphere and reentry into the atmosphere of spacecraft vehicles. The subcommittee recommended to NACA that research in these matters be intensified.
    Minutes of Meeting, NACA Committee on Aerodynamics, Nov. 18-20, 1957, pp. 4-5.

November 19

Preston R. Bassett of the NACA Committee on Aerodynamics presented a resolution urging NACA to adopt an aggressive program in space research technology.

Minutes of Meeting, NACA Committee on Aircraft, Missile and Spacecraft Aerodynamics, March 21, 1958, pp. 3-4.

November 21

The National Advisory Committee for Aeronautics established a Special Committee on Space Technology to study and delineate problem areas that must be solved to make space flight a practical reality and to consider and recommend means for attacking these problems. Dr. H. Guyford Stever of the Massachusetts Institute of Technology was named chairman.
Minutes of Meeting, NACA Committee on Aircraft, Missile and Spacecraft Aerodynamics, March 21, 1958, pp. 3-4; Emme, *Aeronautics and Astronautics: 1915 - 1960*, p. 92.

The Rocket and Satellite Research Panel recommended the creation of a National Space Establishment in the Executive Branch of the Government. According to the proposal, activities of this agency would be under civilian leadership, and the organization would be charged with formulating and supervising a space research program. An annual budget of $1 billion for a period of 10 years was recommended.

House Rpt. 67, 87th Cong., 1st Sess., p. 35. The origin of this particular panel was in 1946, when the V-2 panel was formed of representatives from interested agencies. During its tenure, a total of 60 V-2's were fired. In 1948, the name was changed to Upper Atmosphere Rocket Research Panel and, finally, in 1957 it was redesignated Rocket and Satellite Research Panel.

November 21-22

Over one-half of the NACA Propulsion Conference was devoted to the discussion of possible space propulsion systems. Three particular systems appeared to afford excellent choices for such purposes. These were: the chemical rocket, the nuclear rocket, and the nuclear-electric rocket. It was the considered opinion of the conference members that the chemical rocket would be quite adequate for a round trip to the moon.
Study, *NACA Research into Space*, Dec. 1957.

November (during the month)

A presentation on manned orbital flight was made by Maxime A. Faget. The concept included the use of existing ballistic missiles for propulsion, solid-fuel retrorockets for reentry initiation, and a nonlifting ballistic shape for the reentering capsule. This concept was considered to be the quickest and safest approach for initial manned flights into orbit.
Information supplied by Maxime A. Faget, July 9, 1963.

December 4

The American Rocket Society's proposal for an Astronautical Research and Development Agency, formally presented to President Eisenhower on October 14, 1957, was publicly announced.
    House Rpt. 67, 87th Cong., 1st Sess., p. 36.

December 5

An announcement was made that an Advanced Research Projects Agency would be created in the Department of Defense to direct its space projects.
    House Rpt. 67, 87th Cong., 1st Sess., p. 36.

December 6

IGY Vanguard (TV-3), the first with three live stages, failed to launch a test satellite.
    Emme, *Aeronautics and Astronautics: 1915 - 1960*, p. 92.

December 10

The Air Force created a Directorate of Astronautics to manage and coordinate Astronautical research programs, including work on satellites and antimissile-missile weapons. Brigadier General Homer A. Boushey was named to head the office. Later in the month the order was rescinded by James H. Douglas, Secretary of the Air Force, who considered the creation of such a group before the activation of the Advanced Research Projects Agency to be premature.
    House Rpt. 67, 87th Cong., 1st Sess., p. 36.

PART I (B)
Major Events Leading to Project Mercury
January 1958 through October 1, 1958

1958

January 4

The American Rocket Society and the Rocket and Satellite Research Panel issued a summary of their proposals for a National Space Establishment. The consensus was that the new agency should be independent of the Department of Defense and not, in any event, under one of the military services.
> Emme, *Aeronautics and Astronautics: 1915 - 1960*, p. 94.

January 10

A successful limited flight was made by the fourth Atlas fired from Cape Canaveral.
> House Rpt. 67, 87th Cong., 1st Sess., p. 36.

January 12

President Eisenhower, answering a December 10, 1957, letter from Soviet Premier Nikolai A. Bulganin regarding a summit conference on disarmament, proposed that Russia and the United States ". . . agree that outer space should be used for peaceful purposes." This proposal was compared dedicate atomic energy to peaceful uses, an offer which The Soviets rejected.
> House Document No. 71, 86th Congress, 1st Sess., p. 18; Emme, *Aeronautics and Astronautics: 1915 - 1960*, p. 94.

January 15

The Air Force received 11 unsolicited industry proposals for Project 7969, and technical evaluation was started. Observers from NACA participated. (See March 1956 entry.)
> "Outline of History of USAF Man-in-Space R&D Program," *Missiles and Rockets*, Vol. 10, No. 13 (March 26, 1962), pp. 148-149.

January 16

A resolution was adopted by NACA stating that NACA had an important responsibility for coordinating and conducting research in space technology, either in its own laboratories or by contract. (See November 19, 1957, entry.)

NACA Resolution on the Subject of Space Flight adopted Jan. 16, 1958, contained in NACA Study, *A Program for Expansion of NACA Research in Space Flight Technology with Estimates of Staff and Facilities Required,* Feb. 10, 1958.

Paul E. Purser and Maxime A. Faget conceived of a solid-fuel launch vehicle design for the research and development phase of a manned satellite vehicle project. This launch vehicle was later designated Little Joe. When Project Mercury began in October 1958, the purpose of the Little Joe phase was to propel a full-scale, full-weight developmental version of the manned spacecraft to some of the flight conditions that would be encountered during exit from the atmosphere on an orbital mission. Also, Little Joe tests were used to perfect the escape maneuver in the event of an aborted mission.

Letter, Space Task Group to AVCO-Everett Research Laboratory (no subject), May 5, 1960.

January 29-31

A conference was held at Wright-Patterson Air Force Base, Ohio, to review concepts for manned orbital vehicles. The NACA informally presented two concepts then under study at Langley Aeronautical Laboratory: the one proposed by Maxime A. Faget involved a ballistic, high-drag capsule with heat shield on which the pilot lies prone during reentry, with reentry being accomplished by reverse thrust at the apogee of the elliptical orbit involving a deceleration load of about 8g, and proceeding to impact by a parachute landing; the other Langley proposal called for the development of a triangular platform vehicle with a flat bottom having some lift during reentry. At this same meeting there were several Air Force contractor presentations. These were as follows: *Northrop*, boost-glide buildup to orbital speed; *Martin*, zero-lift vehicle launched by a Titan with controlled flight estimated to be possible by mid-1961; *McDonnell*, ballistic vehicle resembling Faget's proposal, weighing 2,400 pounds and launched by an Atlas with a Polaris second stage; *Lockheed*, a 20 degree semi apex angle cone with a hemispherical tip of 1-foot radius, pilot in sitting position facing rearward, to be launched by an Atlas-Hustler combination; *Convair* reviewed a previous proposal for a large-scale manned space station, but stated a minimum vehicle - a 1,000-pound sphere - could be launched by an Atlas within a year; *Aeronutronics*, cone-shaped vehicle with spherical tip of 1-foot radius, with man enclosed in sphere inside vehicle and rotated to line the pilot up with accelerations, and launched by one of several two-stage vehicles; *Republic*, the Ferri sled vehicle, a 4,000 pound, triangular plan with a two-foot diameter tube running continuous around the leading and trailing edge and serving as a fuel tank for final-stage, solid-propellant rockets located in each wing tip, with a man in small compartment on top side, and with a heat-transfer ring in the front of the nose for a glide reentry of 3,600 miles per hour with pilot ejecting from capsule and parachuting down, and the launch vehicle comprising three stages (also see July 31, 1958 entry); *AVCO*, a 1,500-pound vehicle sphere launched by a Titan, equipped with a stainless-steel-cloth parachute whose diameter would be controlled by compressed air bellows and which would orient the vehicle in orbit, provide deceleration for reentry, and control drag during reentry; *Bell*, reviewed proposals for boost-glide vehicles, but considered briefly a minimum vehicle, spherical in shape, weighing about 3,000 pounds; *Goodyear*, a spherical vehicle with a rearward facing tail cone and ablative surface, with flaps deflected from the cone during reentry for increased drag and control, and launched by an Atlas or a Titan plus a Vanguard second stage; *North American*, extend the X-15 program by using the X-15 with a three-stage launch vehicle to achieve a single orbit with an apogee of 400,000 feet and a perigee of 250,000, range about 500 to 600 miles and landing in the Gulf of Mexico, and the pilot ejecting and landing by parachute with the aircraft being lost.

    Memo, Clarence A. Syvertson to Director, Langley Aeronautical Laboratory, subject: Visit to WADC, Wright-Patterson AFB, Ohio, to Attend Conference on January 29-31, 1958, Concerning Research Problems Associated with Placing a Man in a Satellite Vehicle, Moffett Field, Feb. 18, 1958.

January 31

An Army Jupiter-C missile boosted Explorer I, America's first artificial earth satellite, into orbit. Other than the achievement of orbital conditions, one of the more significant contributions of this flight was the discovery of the Van Allen Radiation Belt, named for Dr. James A. Van Allen, head of the physics department at the State University of Iowa.

    Akens, *Origins of MSFC*, p. 47.

Lieutenant General Donald Putt, Air Force Director of Research and Development, sent a letter to Dr. Hugh Dryden, Director of NACA, inviting NACA participation in the Air Force effort in the manned ballistic rocket program. Dr. Dryden informed the Air Force that NACA was preparing manned spacecraft designs for submission in March 1958.

    Letter, Lt. Gen. D. L. Putt, DSC/Development, Hq. USAF, to Dr. H. L. Dryden, Director, NACA, Jan. 31, 1958.

February 6

The Senate passed a resolution (S Res 256) creating a special Committee on Space and Astronautics to frame legislation for a national program for space exploration.

    Emme, *Aeronautics and Astronautics: 1915 - 1960*, p. 95.

February 7

The Secretary of Defense issued a directive establishing the Advanced Research Projects Agency, an organization under consideration since November 15, 1957. It was to be a centralized group capable of handling direction of both outer space and antimissile-missile projects, whose duties in the space field were to bridge the gap until Congress could consider legislative proposals for the establishment of a National Space Agency.

    House Rpt. 1228, 86th Cong., 2nd Sess., p. 3.

February 10

A study entitled, "A Program for Expansion of NACA Research Space Flight Technology with Estimates of the Staff and Facilities Required" was published by the NACA staff. The study pointed out the urgent need for a rapid buildup of a national capability in space technology leading to early flights of manned space vehicles. Besides devoting some of its laboratory facilities, NACA would integrate into the program the talent and competence of qualified scientific groups outside its organization by a greatly expanded program of contracted research. To support a program of this scope, NACA estimated an additional annual budget of $100 million and 9,000 additional personnel were required. It was also recommended that over the next five years (1958 - 1962) $55 million be expended in new facility construction to support space research projects. In regard to the contracted research facet of the proposal, NACA estimated $10 million a year would be needed at the outset of the program. Besides these recommendations, NACA reviewed the following specific research projects for active consideration: space propulsion systems for launching and flight; materials and structures; space flight research involving launching, rendezvous, reentry, recovery, flight simulation, navigation, guidance, and control; space mechanics and communications; and space environment.

    NACA Study, *A Program for Expansion of NACA Research in Space Flight Technology*, Feb. 10, 1958.

February 13

The Special Committee on Space Technology, established by NACA on November 21, 1957, to study and delineate problem areas that must be resolved to make space flight a practical reality and to consider recommended means for attacking these problems, met for the first time. At the meeting the new committee established seven working groups: (1) objectives, (2) vehicular program, (3) reentry, (4) range, launch, and tracking facilities, (5) instrumentation, (6) space surveillance, and (7) human factors and training. The objectives group was to draft a complete national program for space research. Other than this specific assignment, the remainder of the meeting was largely devoted to organizing the working groups. These groups were to present their first reports at the next meeting.

    Minutes of Meeting, Committee on Aircraft, Missile and Spacecraft Aerodynamics, Mar. 21, 1958, p. 5.

February 14

A report entitled, "Basic Objectives of a Continuing Program of Scientific Research in Outer Space," was presented by the IGY Committee. The committee

was of the opinion that the need for space research would be required far past the close of the IGY in December 1958.
>Emme, *Aeronautics and Astronautics: 1915 - 1960*, p. 95.

February 20

The name of the NACA Committee on Aerodynamics was changed to Committee on Aircraft, Missile, and Spacecraft Aerodynamics to indicate clearly the committee's cognizance over problems applicable to spacecraft and missiles as well as aircraft. The Aerodynamics Committee had been studying spacecraft research problems for the past 6 years.
>Minutes of Meeting, Committee on Aircraft, Missile and Spacecraft Aerodynamics, Mar. 21, 1958, p. 2.

February 27

Experience with the X-15 design indicated that many of the weight figures advanced by the Langley Aeronautical Laboratory for the drag or lift configurations of the reentry vehicle (later to become the Mercury spacecraft) were too low, according to Walter C. Williams, Chief of the NACA High-Speed Flight Station. Weights of auxiliary-power fuel, research instrumentation, and cockpit equipment as set by Langley were too low in terms of X-15 experience. Williams stated the total weight should be 2,300 pounds for the drag configuration and 2,500 pounds for the lifting configuration.
>Letter, NACA Hq. to Langley, subject: Comments on Suggested Ground Rules for Satellite Reentry Vehicles, Feb. 27, 1958.

March 10

Reports were made on recoverable manned satellite configurations being considered by NACA. One involved a blunt, high-drag, zero-lift vehicle that would depend on a parachute landing for final deceleration. Another was a winged vehicle that would glide to a landing after reentering the atmosphere. The third proposal involved features of each of the above. Besides the configuration studies, significant reports were completed relative to motion and heating, stabilization, and attitude control.
>Study, "Satellite and Spacecraft," *Current NACA Aerodynamic Research Relating to Upper Atmosphere and Space Technology*, Mar. 10, 1958, p. 15.

March 10-12

A working conference in support of the Air Force "Man-in-Space Soonest" (MISS) was held at the Air Force Ballistic Missile Division in Los Angeles, California. General Bernard Schriever, opening the conference, stated that events were moving faster than expected. By this statement he meant that Roy Johnson, the new head of the Advanced Research Projects Agency, had asked the Air Force to report to him on its approach to putting a man in space soonest. Johnson indicated that the Air Force would be assigned the task, and the purpose of the conference was to produce a rough-draft proposal. At that time the Air Force concept consisted of three stages: a high-drag, no-lift, blunt-shaped spacecraft to get man in space soonest, with landing to be accomplished by a parachute; a more sophisticated approach by possibly employing a lifting vehicle or one with a modified drag; and a long-range program that might end in a space station or a trip to the moon.

> Memo, Lawrence A. Clousing to Director, Ames Aeronautical Laboratory, subject: Working Conference for the Air Force "Man in Space Soonest" Program, held Mar. 10-12, 1958, at the Air Force Ballistic Missile Division Offices, Los Angeles, Mar. 24, 1958.

March 12

The NACA staff completed a program outline for conducting the manned satellite program. At that time, NACA was already actively engaged in research and study of several phases. For example, in the basic studies category effort had been expended on the study of orbits and orbit control, space physical characteristics, configuration studies, propulsion system research, human factors, structures and materials, satellite instrumentation, range requirements, and noise and vibration during reentry and exit. In addition, NACA outlined the complete program covering full-scale studies of mockups, simulators, and detail designs; full-scale vertical and orbiting flights involving unmanned, animal, and manned flights and recovery; and exploitation of the program to increase the payloads. As to the design concepts for such a program, NACA believed that the Atlas launch vehicle was adequate to meet launch-vehicle requirements for manned orbital flights; that retrograde and vernier controllable thrust could be used for orbital control; that heat-sink or lighter material could be used against reentry heating; that guidance should be ground programed with provisions for the pilot to make final adjustments; that recovery should be accomplished at sea with parachutes used for letdown; that a network of radar stations should be established to furnish continuous tracking; and that launchings be made from Cape Canaveral. It was estimated that with a simple ballistic shape accelerations would be within tolerable limits for the pilot. Temperature control, oxygen supply, noise, and vibration were considered engineering development problems, which could be solved without any special breakthroughs.

> Outline, *Manned Satellite Program*, prepared by NACA Staff, Mar. 12, 1958.

March 17

The NACA Special Committee on Space Technology held its second meeting at the Ames Aeronautical Laboratory, and preliminary reports were presented by the committee working groups on objectives and vehicular programs. The committee as a whole was briefed on the work that had been accomplished by the former NACA Committee on Aerodynamics over the past 6 years. It was stated that between 1952 and 1956, approximately 10 percent of NACA's research efforts were applicable directly or indirectly to astronautics. In 1957, the percentage of space flight research rose to 23; and at the time of the meeting, 30 percent of the aerodynamic effort and 20 percent of propulsion research was

applicable to astronautics problems. The committee also heard special papers on research being conducted in fluid mechanics, satellite studies, spacecraft design proposals, boost-glide and hypersonic vehicle studies, and missiles.
> Minutes of Meeting, Committee on Aircraft, Missile and Spacecraft Aerodynamics, March 21, 1958, p. 6.

March 18

An NACA report was published entitled, "Preliminary Studies of Manned Satellites, Wingless Configuration, Non-Lifting," by Maxime A. Faget, Benjamine Garland, and James J. Buglia. Later this document became the basic working paper for the Project Mercury development program, and was reissued as NASA Technical Note D-1254, March 1962.
> Maxime A. Faget, et al, *Preliminary Studies of Manned Satellites, Wingless Configuration, Non-Lifting*, Langley Aeronautical Laboratory, March 18, 1958.

March 18-20

An "NACA Conference on High-Speed Aerodynamics" was held at the Ames Aeronautical Laboratory, Moffett Field, California, to acquaint the military services and industrial contractors interested in aerospace projects with the results of recent research conducted by the NACA laboratories on the subject of space flight. The conference was attended by more than 500 representatives from the NACA, industry, the military services, and other appropriate government agencies. Some 46 technical papers were presented by NACA personnel, and included specific proposals for manned space flight vehicle projects. One of these was presented by Maxime A. Faget. (See March 18, 1958, entry.) Other papers within the category of manned orbital satellites included: "Preliminary Studies of Manned Satellites, Wingless Configuration, Lifting Body" by Thomas J. Wong and others; "Preliminary Studies of Manned Satellites, Winged Configurations" by John V. Becker; "Preliminary Aerodynamic Data Pertinent to Manned Satellite Reentry Configurations" by Jim A. Penland and William O. Armstrong; and "Structural Design Considerations for Boost-Glide and Orbital Reentry Vehicles" by William A. Brooks and others.
> Papers compiled and presented at *NACA Conference on High-Speed Aerodynamics*, Ames Aeronautical Laboratory, Moffett Field, Calif., March 18-20, 1958, pp. ix-xxi, 19-87.

March (during the month)

At the Langley Aeronautical Laboratory, a working committee studied various manned satellite development plans and concluded that a ballistic-entry vehicle launched with an existing intercontinental ballistic missile propulsion system could be utilized for the first manned satellite project.
> Memo, Warren J. North to NASA Administrator, subject: Background of Project Mercury Schedules, Aug. 14, 1960.

Robert R. Gilruth, Clotaire Wood, and Hartley A. Soule of NACA transmitted a document to the Air Research and Development Command, which listed the design concepts NACA believed should be followed to achieve manned orbital flights at the earliest possible date. These were: (1) design and develop a simple ballistic vehicle, (2) use existing intercontinental ballistic missile propulsion systems, and (3) use the heat sink method for reentry from orbital conditions.

> Memo, Clotaire Wood to Space Flight Development, subject: "Background on WADC Letter to NASA of October 22, 1958, covering Ablation/Heat Sink Investigation - Manned Reentry," Nov. 7, 1958.

April 2

President Eisenhower submitted to Congress a special message calling for the creation of a special civilian space agency, with NACA serving as a nucleus, to conduct federal aeronautic and space activities.

> Memo, Warren J. North to NASA Administrator, subject: Background of Project Mercury Schedules, Aug. 14, 1960.

April (during the month)

Maxime A. Faget and associates conceived the idea of using a contour couch to withstand the high g-loads attendant to acceleration and reentry forces of manned space flight. Fabrication of test-model contour couches was started in the Langley shops in May 1958, and the concept was proved feasible on July 30 (see entry) of that same year.

> Information supplied by Jack C. Heberlig, Engineering and Development, Manned Spacecraft Center, May 23, 1963.

June 5

After serving as a liaison officer of NACA and as a participating member of an Advanced Research Projects Agency panel, Maxime A. Faget reported to Dr. Hugh Dryden on resulting studies and attending recommendations on the subject of manned space flight. He stated that the Advanced Research Projects Agency panel was quite aware that the responsibility for such a program might be placed with the soon-to-be-created civilian space agency, although they recommended program management be placed with the Air Force under executive control of NACA and the Advanced Research Projects Agency. The panel also recommended that the program start immediately even though the specific manager was, as yet, unassigned. Several of the proposals put forth by the panel on the proposed development were rather similar to the subsequent evolvement. The system suggested by the Advanced Research Projects Agency was to be based on the use of the Atlas launch vehicle with the Atlas-Sentry system serving as backup; retrorockets were to be used to initiate the return from orbit; the spacecraft was to be nonlifting, ballistic type, and the crew was to be selected from qualified volunteers in the Army, Navy, and Air Force.

> Memo, Maxime A. Faget to Dr. Dryden, Director, NACA, June 5, 1958.

June 22

NACA personnel discussed the proposed space agency budget, including the manned satellite project, with Bureau of Budget officials.
> Memo, Warren J. North to NASA Administrator, subject: Background of Project Mercury Schedules, Aug. 14, 1960.

June 26

Meetings were held with NACA, AVCO, and Lockheed representatives in attendance to consider materials for thermal protection of satellite reentry vehicles.
> Memo, H. M. Henneberry and G. C. Deutsch to Associate Director, NASA-Langley, subject: Discussions with AVCO and Lockheed Representatives Concerning Materials for Thermal Protection of Satellite Reentry Vehicles, Washington D.C., June 26-27, 1958, Sept. 8, 1958.

June (during the month)

Preliminary specifications of the first manned satellite vehicle were drafted by Langley Aeronautical Laboratory personnel under the supervision of Maxime Faget and Charles W. Mathews. After a number of revisions and additions, these specifications were used for the Project Mercury spacecraft contract with McDonnell Aircraft Corporation. A working group of representatives from the Langley Aeronautical Laboratory and the Lewis Flight Propulsion Laboratory was formed for the purpose of outlining a manned satellite program.
> Information supplied by Maxime A. Faget, July 9, 1963.

NACA representatives were assigned to the Advanced Research Projects Agency, Manned Satellite Committee.

> Memo, Warren J. North to NASA Administrator, subject: Background of Project Mercury Schedules, Aug. 14, 1960.

July 9

General Electric Company personnel presented a briefing at NACA headquarters on studies related to manned space flight. The company held contracts let by the Wright Air Development Center for study and mock-up of a manned spacecraft. NACA made no official comment.
> Memo for Files, Hugh Henneberry, NACA Space Flight Office, subject: Briefing by General Electric Representatives on Studies Related to Man-in-Space Program, July 17, 1958.

July 15

Cook Electric Company submitted a proposal to the McDonnell Aircraft Corporation as a part of a preliminary study and design effort by McDonnell for a manned satellite. McDonnell, prior to being awarded the Mercury prime development contract in February 1959, spent 11 months under a company research budget working on a manned orbital spacecraft concept.

>Chronological statement filed by Cook Electric Company with NASA Hq., March 3, 1959.

July 16

Congress passed the National Aeronautics and Space Act of 1958.

>Public Law 85-568, 85th Congress, H.R. 12575, subject: National Aeronautics and Space Act of 1958, July 29, 1958.

July 18

In a memorandum to Dr. James R. Killian, Jr., Special Assistant to the President for Science and Technology, Dr. Hugh L. Dryden, Director of NACA, pointed out that NASA would inherit from NACA a rich technical background, competence, and leadership in driving toward the objective of a manned satellite program. For years NACA groups had been involved in research on such items as stabilization of ultra-high speed vehicles, provision of suitable controls, high temperature structural designs, and all the problems of reentry. In fact, a part of this work had been directed specifically toward the problem of designing a manned satellite. Also, the X-15 program had provided much experience in human factors applicable to the orbital flight of man. Therefore, Dr. Dryden concluded, in consonance with the intent of the Space Act of 1958, the assignment of the program to the NACA would be consistent.

>Memo, Dr. H. L. Dryden, Director, NACA, to Dr. J. R. Killian, Jr., subject: Manned Satellite Program, July 18, 1958.

July 29

The National Aeronautic and Space Act of 1958 was signed into a law by President Eisenhower.

>Public Law 85-568, 85th Congress, H.R. 12575, subject: National Aeronautics and Space Act of 1958, July 29, 1958.

July 30

By using the development model of the Mercury contour couch designed by Maxime A. Faget and associates, Carter C. Collins withstood a 20g load on the centrifuge at Johnsville, Pennsylvania. This test proved that the reentry accelerations of manned space flight could be withstood.

>Information supplied by Maxime A Faget, Assistant Director for Engineering and Development, MSC.

July 31

Republic Aviation representatives briefed NACA Headquarters personnel on the man-in-space studies in which the company had been engaged since the first of

the year. They envisioned a four-stage solid launch vehicle system and a lifting reentry vehicle, which was termed a sled. The vehicle was to be of triangular shape with a 75 degree leading-edge sweep. Aerodynamic and reaction controls would be available to the pilot. For the launch vehicle, Republic proposed a Minuteman first stage, a Polaris first stage, a Minuteman upper stage, and a Jumbo rocket fourth stage. Other details relative to reentry and recovery were included in the briefing.
    Memo, Hugh M. Henneberry, NACA Lewis, to the files, subject: Briefing by Republic Aviation Representatives on Man-in-Space Studies, Aug. 5, 1958.

July (during the month)

The initial concept of the use of a tractor rocket for an escape device was suggested by Maxime A. Faget - an idea which developed into the Mercury escape rocket. (see fig. 1.)
    Information supplied by Maxime A. Faget, July 9, 1963.

Figure 1. Closeup view of Recruit escape rocket and full-scale spacecraft.

August 1

Dr. Hugh L. Dryden, NACA Director, presented a program on the technology of manned space flight vehicles to the Select Committees of Congress on Astronautics and Space Exploration.
    House Report No. 671, *Project Mercury, Second Interim Report*, 87th Congress, 1st Session (June 29, 1961), p. 8.

August 8

A memorandum from the Secretary of the Army to the Secretary of Defense recommended Project Adam for a manned space flight program. This plan proposed a ballistic suborbital flight using existing Redstone hardware as a national political-psychological demonstration. This memo proposed that funds in

the amount of $9 million and $2.5 million for fiscal years 1959 and 1960, respectively, be approved for program execution.
> House Rpt. 1228, 86th Cong., 2nd Sess., p. 3; David S. Akens, *History of Marshall Space Flight Center, July 1-Dec. 31, 1960*, Appendix B, "Mercury-Redstone Chronology," p. 3. Hereinafter cited as Akens, *History of MSFC*, Mercury-Redstone Chronology.

August (during the month)

President Eisenhower assigned the responsibility for the development and execution of a manned space flight program to the National Aeronautics and Space Administration. However, NASA did not become operational until October 1, 1958.
> House Rpt. 671, 87th Cong., 1st Sess., p. 8.

September 11

At an Army Advanced Research Projects Agency conference, the Army was advised there was little chance for approval of Project Adam.
> Akens, *History of MSFC*, Mercury-Redstone Chronology, p. 4.

September 17

A joint National Aeronautics and Space Administration/Advanced Research Projects Agency Manned Satellite Panel was formed. This panel, with the aid of technical studies prepared by the Langley and Lewis Research Centers and assistance from the military services, drafted specific plans for a program of research leading to manned space flight.
> Emme, *Aeronautics and Astronautics: 1915 - 1960*, p. 102.

September 25

Dr. T. Keith Glennan, NASA Administrator, announced publicly that NASA would be activated on October 1, 1958.
> Emme, *Aeronautics and Astronautics: 1915 - 1960*, p. 102.

September (during the month)

Study was started on the tracking and ground instrumentation networks for the manned satellite project.
> NASA Space Task Group, *Project Mercury [Quarterly] Status Report No. 1 for Period Ending January 31, 1959*.

September 24-October 1

A series of meetings were held in Washington, with Robert R. Gilruth serving as chairman to draft a manned satellite program and provide a basic plan for meeting the objectives of this program. Others attending included S. B. Batdorf,

A. J. Eggers, Maxime A. Faget, George Low, Warren North, Walter C. Williams, and Robert C. Youngquist.

> NASA Minutes of Meeting, subject: Panel for Manned Space Flight, Sept. 24, 30 and Oct. 1, 1959.

October 1

NASA was activated in accordance with the terms of Public Law 85-568, and the nonmilitary space projects which had been conducted by the Advanced Research Projects Agency were transferred to the jurisdiction of the NASA. Concurrently, NACA, after a 43-year tenure, was inactivated, and its facilities and personnel became a part of NASA.

> House Rpt. 67, 87th Cong., 1st Sess., p. 57.

1958 (during the year)

The Navy space proposal to the Advanced Research Projects Agency, during the tenure of that organization's interim surveillance over national space projects, was known as Project Mer. This plan involved sending a man into orbit in a collapsible pneumatic glider. The glider and its occupant would be launched in the nose of a giant launch vehicle. After the glider had been placed in orbit, it would be inflated, and then flown down to a water landing.

> House Rpt. 1228, 86th Cong., 2nd Sess., p. 4.

PART II (A)
Research and Development Phase of Project Mercury

October 3, 1958 through December 1959

1958

October 3-7

Studies and plans of the manned satellite project were presented to Advanced Research Projects Agency on October 3 and to Dr. T. Keith Glennan, NASA Administrator, on October 7. On October 7, 1958, Dr. Glennan approved the project by saying, in effect, "Let's get on with it." (See fig. 2)
> Memo, Warren J. North to NASA Administrator, subject: Background of Project Mercury Schedules, Aug. 14, 1960.

Figure 2. Mercury spacecraft in orbit: Artist's conception.

October 6

Personnel from the Langley Research Center visited the Army Ballistic Missile Agency to open negotiations for procuring Redstone and Jupiter launch vehicles for manned satellite projects.
> Memo, Space Task Group to NASA Hq., subject: Transmittal of Comments on AOMC Memorandum for Record - Meeting of NASA and AOMC, Oct. 6, 1958, Nov. 13, 1958, with enclosures.

October 7-8

Personnel from the Space Task Group involved in the study of reentry methods visited the Air Force Wright Air Development Center, Dayton, Ohio, for the purpose of preparing test specimens. Along with individuals from the center and the Air Force Ballistic Missile Division, the group then met at the Chicago Midway Laboratories, Chicago, Illinois, to investigate various ablation methods of reentry. Concurrently, these same methods were being investigated at high-temperature test facilities at Langley.
        Letter, Wright Air Development Center to Air Research and Development Command, subject: Ablation/Heat Sink Investigation - Manned Reentry, Oct. 21, 1958.

October 9

In behalf of the manned satellite project, an air drop program for full-scale parachute and landing system development was started at Langley.
        NASA Space Task Group, *Project Mercury [Quarterly] Status Report No. 1 for Period ending January 31, 1959,* March 1959.

October 14

The Assistant Secretary of Defense for Supply and Logistics invited the National Aeronautics and Space Administration to submit nominations for materiel procurement urgency (commonly known as the DX priority rating).
        Notes, Assistant to Deputy Administrator to NASA Administrator, subject: Briefing Memorandum for the Administrator, March 12, 1959.

October 17-18

Langley Research Center personnel visited the Air Force Ballistic Missile Division, Inglewood, California, to open negotiations for procuring Atlas launch vehicles for the manned satellite project.
        Memo, Warren J. North to NASA Administrator, subject: Background of Project Mercury Schedules, Aug. 14, 1960.

October 21

A bidders' briefing for the Little Joe launch vehicle was held. As earlier mentioned, this launch vehicle was to be used in the development phase of the manned satellite project. (See January 16, 1958, entry). The Little Joe launch vehicle was 48 feet in height, weighed (at maximum) 41,330 pounds, was 6.66 feet in diameter, consisted of four Pollux and four Recruit clustered, solid-fuel rockets, could develop a thrust of 250,000 pounds, and could lift a maximum payload of 3,942 pounds. (See fig. 3.)
        Letter, Space Task Group to AVCO-Everett Research Laboratory, (no subject), May 5, 1960.

Figure 3. Little Joe on launcher at Wallops during checkout.

October 23

Preliminary specifications for a manned spacecraft were established to industry. These specifications outlined the program and suggested methods of analysis and construction.

> Memo, George Low, NASA, for the House Committee on Space and Astronautics, subject: NASA Procedure Used in the Selection of McDonnell Corporation for the Construction of the Project Mercury Capsule, April 24, 1959.

October 27

A special Committee on Life Sciences was established at Langley to determine qualifications and attributes required of personnel to be selected for America's first manned space flight and to give advice on other human aspects of the manned satellite program.

> Letter, Charles J. Donlan, Associate Director, STG, to Willson H. Hunter, NASA Headquarters, subject: Transmittal of Materials Requested by Willson H. Hunter, Dec. 16, 1960. This letter outlines the overall sequence of events in the astronaut selection program.

October (during the month)

Drop tests of full-scale capsules from a C-130 airplane were started to check parachute deployment and spacecraft stability. Preliminary drops of the parachute system were made from a NASA helicopter at West Point, Virginia.

These drops involved the use of a concrete-filled drum attached to an operating canister system. The purpose of this phase was to demonstrate the adequacy of the mechanical system of deploying the parachutes. Subsequently, the drops were made by the C-130's at Pope Field, North Carolina, from low levels to perfect a means of extracting the spacecraft from the aircraft. Full-scale spacecraft and operating parachutes were used in these drops, and all operational features of the drop-test program were worked out. The next phase was the research and development drops offshore of Wallops Island, Virginia, and the objectives here were as follows: to study the stability of the spacecraft during free fall and with parachute support; to study the shock input to the spacecraft by parachute deployment; and to study and develop retrieving operations.

>Memo, George M. Low to NASA Administrator, subject: Status of Manned Satellite Project, November 25, 1958.

Design work was started on the Little Joe vehicles and test model spacecraft.

>Memo, George M. Low to NASA Administrator, subject: Status of Manned Satellite Project, November 25, 1958.

Dr. W. Randolph Lovelace II was appointed by NASA Headquarters as Chairman of a Special Committee on Life Sciences by T. Keith Glennan, the NASA Administrator. After prospective astronaut candidates were interviewed in Washington, D. C., those chosen for further consideration received medical examinations at the Lovelace Clinic in Albuquerque, New Mexico (see Feb. 1-14, 1959, entry).

>House Document No. 454, 86th Cong., 2nd Sess., subject: Third Semiannual Report of the National Aeronautics and Space Administration, Message from the President of the United States, Aug. 30, 1960.

November 3

The initial contingent of military service aeromedical personnel reported for duty and began working on human factors, crew selection, and crew training plans for the manned spacecraft program.

>NASA Space Task Group, *Project Mercury [Quarterly] Status Report No. 1, for Period Ending January 31, 1959.*

November 5

The Space Task Group, unofficially established on October 8, 1958, was officially formed at Langley Field, Virginia, to implement a manned satellite project. Robert R. Gilruth and Charles J. Donlan were appointed as Project Manager and Assistant Project Manager, respectively. The memorandum of establishment listed a total of 35 people from Langley assigned to the Space Task Group. The following personnel were transferred from the Langley Research Center to the newly established Space Task Group: *Robert R. Gilruth, Charles J. Donlan,

*Paul E. Purser, *Maxime A. Faget, Charles H. Zimmerman, *William M. Bland, *Aleck C. Bond, Alan B. Kehlet, *Charles W. Mathews, *Edison M. Fields, *Robert G. Chilton, *Jerome B. Hammack, *Jack C. Heberlig, *Claiborne R. Hicks, Ronald Kolenkiewicz, *Christopher C. Kraft, *Howard C. Kyle, *William T. Lauten, *John B. Lee, *George F. MacDougall, *John P. Mayer, *William C. Muhley, *Herbert G. Patterson, Harry H. Ricker, Frank C. Robert, Joseph J. Rollins, Ronelda F. Sartor, Paul D. Taylor, Shirley J. Hartley, Norma L. Livesay, Betsy F. Magin, Jacquelyn B. Stearn, *Julia R. Watkins, *Nancy C. Lowe, and Shirley P. Watkins. (* Assigned to Manned Spacecraft Center as of November 1962). Personnel detailed from the Lewis Research Center to the Space Task Group and Project Mercury were as follows: E. H. Buller, A. M. Busch, W. R. Dennis, M. J. Krasnican, *Glynn S. Lunney, *Andre J. Meyer, W. R. Meyer, W. J. Nesbitt, *Gerald J. Pesman, and Leonard Rabb. Individuals from Lewis remained on a detailed status until 1959 when they were permanently reassigned to the Space Task Group. The 45 people listed above were the embryo work force of Project Mercury.

As a note of interest, on the fourth anniversary of the activation document, 21 of the original Langley reassignees and 3 of the Lewis group were members of the Manned Spacecraft Center, the successor of the Space Task Group.

> Memo, Floyd L. Thompson, Acting Director, NASA Langley to all concerned, subject: Space Task Group, Nov. 5, 1958; information supplied by Lynn Manley, Lewis Research Center, May 28, 1963.

November 7

A contractor briefing, attended by some 40 prospective bidders on the manned spacecraft, was held at the Langley Research Center. More detailed specifications were then prepared and distributed to about 20 manufacturers who had stated an intention to bid on the project.

> Memo, George Low to NASA Administrator, subject: Status Report No. 1, Manned Satellite Project, Dec. 9, 1958; Agenda for Prospective Bidders for Manned Satellite Capsule, prepared by Space Task Group for Nov. 7, 1958.

November 14

Specifications for the manned spacecraft (Specification Number S-6) were issued, and final copies were mailed on November 17, 1958, to 20 firms which had indicated a desire to be considered as bidders.

> Memo, Abe Silverstein to NASA Administrator, subject: Schedule for Evaluation and Contractual Negotiations for Manned Satellite Capsule, Dec. 24, 1958; NASA-Langley, subject: Specifications for Manned Space Capsule, Nov. 14, 1958.

The highest national procurement priority rating (DX) was requested for the manned spacecraft project.

> Letter, Hugh L. Dryden to Robert R. Gilruth, (no subject), March 23, 1959.

Twenty firms notified the National Aeronautics and Space Administration of their intention to prepare proposals for the development of the manned spacecraft. NASA set the deadline for proposal submission as December 11, 1958.

> Memo, George M. Low to NASA Administrator, subject: Status Report No. 1, Manned Satellite Project, Dec. 9, 1958; Memo, Robert R. Gilruth to all Space Task Group Personnel, subject: Prime Bidders for Manned Satellite Capsule, Nov. 19, 1958.

November 20

The three military services were invited to send one man each to the Space Task Group to perform liaison duties for the manned spacecraft project. These posts were filled in January 1959 by Lt. Colonel Martin Raines, Army; Lt. Colonel Keith Lindell, Air Force; and Commander Paul Havenstein, Navy.

> Memo, George M. Low to NASA Administrator, subject: Status Report No. 2, Manned Satellite Project, Dec. 17, 1958.

November 24

The Space Task Group placed an order for one Atlas launch vehicle with the Air Force Missile Division, Inglewood, California, as part of a preliminary research program leading to manned space flight. The National Aeronautics and Space Administration Headquarters requested that the Air Force construct and launch one Atlas C launch vehicle to check the aerodynamics of the spacecraft. It was the intention to launch this missile about May 1959 in a ballistic trajectory. This was to be the launch vehicle for the Big Joe reentry test shot, but plans were later changed and an Atlas Model D launch vehicle was used instead.

> Message, NASA NDA, Ralph Cushman, Contracting Officer, NASA, to Commanding General, Air Force Ballistic Missile Division, Nov. 24, 1958.

November 26

The manned satellite program was officially designated Project Mercury.
> Emme, *Aeronautics and Astronautics: 1915-1960*, p. 104.

Space Task Group personnel presented a proposed program for Langley Research Center support in the Little Joe phase of Project Mercury. Langley was favorably inclined, and after a survey of manpower and facility availability, notified Space Task Group on December 5, 1958, of its willingness to support the program. Langley tasks involved contracting for engineering, construction, services, data processing, analysis, and reporting research results.

> Memo, Carl A. Sandahl for Associate Director, NASA Langley, subject: NASA Participation in Little Joe Project, Dec. 9, 1958.

November 28

Less than 18 months after the first flight, an Atlas launch vehicle was launched 6,300 miles down range from Cape Canaveral, Florida.
> House Rpt. 67, 87th Cong. 1st Sess., March 8, 1961.

November (during the month)

A scale model of the Mercury spacecraft (without escape tower), oriented for the reentry phase, was tested at transonic Mach numbers in a 1-foot transonic test tunnel at the Arnold Engineering Development Center, Tullahoma, Tennessee.
> Notes supplied by Marvin E. Hintz, Historical Office, Arnold Engineering Development Center, Tullahoma, Tenn.

November-December

Study was started on spacecraft recovery operations. During this study period, it was learned that the retrieving operation could be very difficult; but with properly designed equipment, helicopter pickup could be used and appeared to be the most favorable method.
> Memo, Warren J. North to NASA Administrator, subject: Background of Project Mercury Schedules, Aug. 14, 1960.

December 1

Design of the Big Joe spacecraft for the Project Mercury reentry test (the spacecraft would be boosted by an Atlas launch vehicle over a ballistic trajectory) was accomplished by the Space Task Group. Construction of the spacecraft was assigned as a joint task of the Langley and Lewis Research Centers under the direction of the Space Task Group. The instrument package was developed by Lewis personnel assigned to the Space Task Group, and these individuals later became the nucleus of the Space Task Group's Flight Operations Division at Cape Canaveral.
> Memo, Warren J. North to NASA Administrator, subject: Background of Project Mercury Schedules, August 14, 1960; information supplied by Aleck Bond, Manned Spacecraft Center, June 11, 1963.

December 2

Space Task Group officials visited the Army Ballistic Missile Agency to determine the feasibility of using the Jupiter launch vehicle for the intermediate phase of Project Mercury, to discuss the Redstone program, and to discuss the cost for Redstone and Jupiter launch vehicles.
> Memo, Warren J. North to Assistant Director for Advanced Technology, subject: Visit to ABMA Regarding Boosters for Manned Satellite and Juno II Programs, Dec. 4, 1958.

December 8

The Space Task Group indicated that nine Atlas launch vehicles were required in support of the Project Mercury manned and unmanned flights and these were ordered from the Air Force Ballistic Missile Division.
> Memo, Warren J. North to NASA Administrator, subject: Background of Project Mercury Schedules, Aug. 14, 1960.

December 9

An aeromedical selection team composed of Major Stanley C. White, Air Force; Lt. Robert B. Voas, Navy; and Captain William Augerson, Army, drafted a tentative astronaut selection procedure. According to the plan, representatives from the services and industry would nominate 150 men by January 21, 1959; 36 of these would be selected for further testing which would reduce the group to 12; and in a 9-month training period, a hard core of 6 men would remain. At the end of December 1958, this plan was rejected.
> Memo, George Low to NASA Administrator, subject: Status Report No. 1, Manned Satellite Project, Dec. 9, 1958; Memo, George Low to NASA Administrator, subject: Status Report No. 3, Project Mercury, Dec. 27, 1958.

December 10

The Space Task Group appointed a Technical Assessment Committee, with Charles H. Zimmerman serving as chairman, to assist the National Aeronautics and Space Administration Source Selection Board. This group provided the board with technical ratings on contractor proposals. Technical specialists throughout the Space Task Group supplied specific component assessment information to the committee.
> Memo, Robert R. Gilruth, subject: Procedures for Technical Assessment of Manufacturer's Proposals for a Manned Space Capsule Submitted in Response to Requests for Proposals on Specification S-6, Nov. 14, 1958, Dec. 10, 1958.

December 11

The Lewis Research Center presented its funding requirements for the attitude control and instrumentation systems for the Big Joe flight test spacecraft. Confirmation of agreements and fund transfer were forwarded by the Space Task Group to Lewis on February 17, 1959.
> Memo, Space Task Group to NASA Headquarters, subject: Request for Transfer of Space Research and Development Funds to Lewis for Manned Space Capsule Instrumentation, Dec. 24, 1958.

Eleven firms submitted proposals for the development of a manned spacecraft. These were AVCO, Chance-Vought, Convair, Douglas, Grumman, Lockheed, Martin, McDonnell, North American, Northrop, and Republic. In addition, Winzen Research Laboratories submitted an incomplete proposal.

Memo, George Low to NASA Administrator, subject: Status Report No. 2, Manned Satellite Project, Dec. 17, 1958.

December 12

Robert R. Gilruth, Mercury Project Manager, requested that the Lewis Flight Research Branch provide technical support for Project Mercury. The Space Task Group was particularly interested in Lewis' instrumentation facilities for use in research and development tests of Big Joe.
> Memo, G. Merritt Preston, Chief, Flight Problems Branch Lewis Research Center, to Dr. Abe Silverstein, NASA Hq., subject: Distribution of Lewis Flying Research Personnel Space Activities, Dec. 12, 1958.

Space Task Group personnel began technical assessment of manned spacecraft development proposals submitted by industry. Charles Zimmermann headed the technical assessment team.

> Memo, George Low to NASA Administrator, subject: Status Report No. 1, Manned Satellite Project, Dec. 9, 1958.

Space Task Group received a "Development and Funding Plan" from the Army Ordnance Missile Command in support of Project Mercury.

> Source as cited.

December 13

Gordo, a primate, was launched into space aboard an Army Jupiter missile nose cone. Although nose cone recovery efforts failed because the float mechanism attached to the nose cone did not function, telemetry data provided useful biomedical information and disclosed that the Navy-trained squirrel monkey had withstood the space flight and reentry phase without any adverse physiological effects. Gordo was in a weightless state for 8.3 minutes, he experienced a 10g pressure in takeoff, and a 40g pressure upon reentry at 10,000 miles per hour.
> Akens, *Origins of MSFC*, Dec. 1960; House Rpt. 67, 87th Cong., 1st Sess., Mar. 8, 1961.

December 17

Dr. T. Keith Glennan referred to the manned satellite project as Project Mercury in a policy speech for the first time.
> Memo, George Low to Dr. Silverstein, NASA Hq., subject: Change of Manned Satellite Name from "Project Mercury" to "Project Astronaut," Dec. 12, 1958.

December 29

A contract was awarded to North American Aviation for design and construction of the Little Joe air frame.
> Memo, George Low to NASA Administrator, subject: Status Report No. 4, Project Mercury, Jan. 12, 1959.

December 30

Space Task Group's technical assessment teams completed the evaluation of industry proposals for design and construction of a manned spacecraft and forwarded their findings to the Source Selection Board, NASA Headquarters.
> Memo, Warren J. North to NASA Administrator, subject: Background of Project Mercury Schedules, August 14, 1960.

December 31

The letter-of-intent was placed with North American Aviation for the fabrication of the Little Joe Test vehicle air frame. Delivery of the air frames for flight testing was scheduled to occur every three weeks beginning in June 1959. Space Task Group had ordered all the major rocket motors, which were scheduled for delivery well ahead of the Little Joe flight test schedule. The spacecraft for this phase of the program was being designed and construction would start shortly. Thus the Little Joe program should meet its intended flight test schedule.
> NASA Space Task Group, *Project Mercury [Quarterly] Status Report No. 1 for Period Ending January 31, 1959*, March 1959.

December (during the month)

A draft checklist entitled "Overall Technical Assessment of Proposals for Manned Space Capsule," was prepared by the Space Task Group for use by the Source Selection Board.
> Source as cited.

1959

January 5

Qualifications were established for pilot selection in a meeting at the NASA Headquarters. These qualifications were as follows: age, less than 40; height, less than 5 feet 11 inches; excellent physical condition; bachelor's degree or equivalent; graduate of test pilot school; 1,500 hours flight time; and a qualified jet pilot.
> Memo, George Low to NASA Administrator, subject: Status Report No. 6, Project Mercury, Feb. 3, 1959.

January 6

A meeting was held at the National Aeronautics and Space Administration Headquarters to discuss the method for spacecraft heat protection. Two plans were considered: beryllium heat sink and ablation. Based on this meeting a decision was made to modify the spacecraft structure in order to accommodate interchangeably ablation heat shields and beryllium heat sinks , and orders were placed for 12 and 6, respectively. The material chosen for the ablation heat was Fiberglas bonded with a modified phenolic resin. This material was found to have good structural properties even after being subjected to reentry heating.
> Memo, George Low to NASA Administrator, subject: Status Report No. 4, Project Mercury, Jan. 12, 1959.

January 9

The Source Selection Board at NASA Headquarters composed of Abe Silverstein, Ralph Cushman, George Low, Walter Schier, DeMarquis Wyatt, and Charles Zimmerman, completed their findings and reported to Dr. T. Keith Glennan, the Administrator. McDonnell Aircraft Corporation was selected as the prime contractor to develop and produce the Mercury spacecraft.
> Memo, George Low to NASA Administrator, subject: Status Report No. 4, Project Mercury, Jan. 12, 1959.

Representatives of the National Aeronautics and Space Administration and the Department of Defense met to coordinate requirements of the two agencies and arrived at an agreement for a "National Program to Meet Satellite and Space Vehicle Tracking Requirements for FY59 and FY60." This meeting led to the formation of a continuing NASA-DOD Space Flight Tracking Resources Committee.

> Emme, *Aeronautics and Astronautics: 1915-1960*, p. 106.

January 14

Preliminary negotiations were started with McDonnell on the technical and legal aspects of the Mercury spacecraft research and development program.
> Memo, George Low to NASA Administrator, subject: Status Report No. 5, Project Mercury, Jan. 20, 1959.

January 16

NASA requested the Army Ordnance Missile Command, Huntsville, Alabama, to construct and launch eight Redstone launch vehicles and two Jupiter launch vehicles in support of Project Mercury manned and unmanned flights.
> Message, NASA Hq. to Commanding General, Army Ordnance Missile Command, Jan. 16, 1959.

During a meeting of the Space Task Group, it was decided to negotiate with McDonnell for design of spacecraft that could be fitted with either a beryllium heat sink or an ablation heat shield. Robert R. Gilruth, the project director, considered that for safety purposes, both should be used. He also felt that the recovery landing bag should be replaced by a honeycombed crushable structure. At this same meeting, a tentative decision was also made that design, development, and contract responsibilities for the Mercury tracking network would be assigned to the Langley Research Center.

> Memo, George Low to NASA Administrator, subject: Status Report No. 5, Project Mercury, Jan. 20, 1959.

January 21

The screening of records for prospective astronauts began.
> Memo, Warren J. North to NASA Administrator, subject: Background of Project Mercury Schedules, Aug. 14, 1960.

January 23

Funds in the amount of $1,556, 200 were made available to the Langley Research Center for the Little Joe development program. The remaining funds of total program costs ($3, 946,000) had already been made available to Langley in a previous transfer of funds.
> Memo, George Low to Dr. Silverstein, subject: Fund Transfer to Langley Research Center for Little Joe Program, Jan. 23, 1959.

January 25

The pilot egress trainer was received from McDonnell and rough water evaluation of the equipment was started immediately by Space Task Group personnel. (See fig. 4.)
> NASA Space Task Group, *Project Mercury [Quarterly] Status Report No. 1 for Period Ending January 31, 1959.*

Figure 4. *Pilot egress trainer.*

January 26

NASA completed contract negotiations with McDonnell for the design and development of the Mercury spacecraft. (See fig. 5.) At that time, McDonnell estimated that the first 3 spacecraft could be delivered in 10 months. Spacecraft refinements slipped this estimated goal by only 2 months.
>Memo, George Low to NASA Administrator, subject: Status Report No. 6, Project Mercury, Feb. 3, 1959.

Figure 5. Manufacture of Mercury spacecraft at McDonnell plant, St. Louis, Mo.

January 29

The Little Joe flight test program was drafted. This plan was updated on April 14, 1959. Primary objectives of the test were to investigate flight dynamics, check drogue parachute operations, determine physiological effects of acceleration on a small primate, and, to some extent, check the spacecraft aerodynamic characteristics.
>NASA Space Task Group, *Project Mercury [Quarterly] Status Report No. 1 for Period Ending January 31, 1959*, March 1959.

January 30

Admiral Arleigh Burke, Chief of Naval Operations, advised Dr. T. Keith Glennan that Navy candidates for Project Mercury had started in the first selection process.
>Letter, Admiral Arleigh Burke, Chief of Naval Operations to Dr. T. Keith Glennan, NASA Administrator (no subject), Jan. 30, 1959.

January (during the month)

McDonnell, as prime contractor, selected Minneapolis-Honeywell as subcontractor for the Mercury stabilization system. At that time, other subcontractors were under consideration for the fabrication of various components: Bell Aircraft Rockets Division, reaction control system; and General Electric, Barnes Instruments, and Detroit Controls were being considered for fabrication of the horizon scanner. Later Bell and Barnes were awarded contracts for respective components.

> NASA Space Task Group, *Project Mercury [Quarterly] Status Report No. 1 for Period Ending January 31, 1959*, March 1959.

Balloon flights were planned for high-altitude qualification tests of the complete spacecraft, including all instrumentation, retrorockets, drogue parachute system, and recovery. Later balloon flights would be manned to provide as much as 24 hours of training followed by recovery at sea. The Space Task Group made surveys of organizations experienced in the balloon field and recommended that the Air Force Cambridge Research Center be given responsibilities for designing, contracting, and conducting the balloon program.

> NASA Space Task Group, *Project Mercury [Quarterly] Status Report No. 1 for Period Ending January 31, 1959*; Memo, George Low to NASA Administrator, subject: Status Report No. 5, Project Mercury, Jan. 20, 1959.

Development of the Mercury pressure suit was started.

> NASA Space Task Group, *Project Mercury [Quarterly] Status Report No. 1 for Period Ending January 31, 1959*, March 1959.

Animal payloads, including pigs and small primates, were planned for some of the Little Joe test flights.

> Memo, George Low to NASA Administrator, subject: Status Report No. 5, Project Mercury, Jan. 20, 1959.

January-February

Study contracts were awarded to Aeronutronics, Space Electronics, and the Massachusetts Institute of Technology Lincoln Laboratory for assistance in developing plans for tracking and ground instrumentation for Project Mercury.
> Memo, Warren J. North to NASA Administrator, subject: Background of Project Mercury Schedules, Aug. 14, 1960.

January-July

Investigations were conducted at the Arnold Engineering Development Center, Tullahoma, Tennessee, in support of Project Mercury. Models of the Mercury

spacecraft were tested at speeds of Mach 8, 16, and 20 to investigate stability, heat transfer, and pressure distribution of Mercury components.
> "A Chronology of the Arnold Engineering Development Center," *AFSC Historical Publications*, Series 62-101. Hereinafter cited as "Chronology of Arnold Development Center."

February 1-14

Some 508 records were reviewed for prospective pilot candidates of which about 110 appeared to qualify. The special committee on Life Sciences decided to divide these into two groups and 69 prospective pilot candidates were briefed and interviewed in Washington. Out of this number, 53 volunteered for the Mercury program, and 32 of the 53 were selected for further testing. The committee agreed there was no further need to brief other individuals, because of the high qualities exhibited in the existing pool of candidates. These 32 were scheduled for physical examination at the Lovelace Clinic, Albuquerque, New Mexico.
> Memo, George Low to NASA Administrator, subject: Status Report No. 5, Project Mercury, February 3, 1959.

February 5

The Navy agreed to perform field service functions in procurement and supply in support of Project Mercury at the McDonnell Aircraft Corporation plant site.
> Letter, NASA Hq. to Chief of Navy Materiel, Department of the Navy (no subject), Apr. 10, 1959.

NASA personnel visited the Wright Air Development Center to investigate its methods and facilities for measuring airborne noise and vibrations.

> Memo, Michael A. Wedding to Lewis Space Task Group, subject: Lewis Space Task Group's visit to Wright Patterson Air Force Base, Dayton, Ohio, on February 5, 1959, Feb. 26, 1959.

February 6

Following industry-wide competition, a formal contract for research and development of the Mercury spacecraft was negotiated with the McDonnell Aircraft Corporation. The contract called for design and construction of 12 Mercury spacecraft, but it did not include details on changes and ground support equipment which were to be negotiated as the project developed. Later, orders were placed with the company for eight additional spacecraft, two procedural trainers, an environmental trainer, and seven checkout trainers. McDonnell had been engaged in studying the development of a manned spacecraft since the NACA presentation in mid-March of 1958.
> Memo, George Low to NASA Administrator, subject: Status Report No. 7, Project Mercury, Feb. 17, 1959.

February 7

At the Lovelace Clinic, Albuquerque, New Mexico, the medical tests for the Mercury astronaut selection were started.
> Memo, George Low to NASA Administrator, subject: Status Report No. 8, Project Mercury, Mar. 4, 1959.

February 10

Wind tunnel tests of Project Mercury configuration models were started. By the end of the year, over 70 different models had been tested by facilities at the Air Force's Arnold Engineering Development Center and the NASA Langley, Ames (fig. 6.), and Lewis Research Centers.
> NASA Space Task Group, *Project Mercury [Quarterly] Status Report No. 5 for Period Ending January 31 1960.*

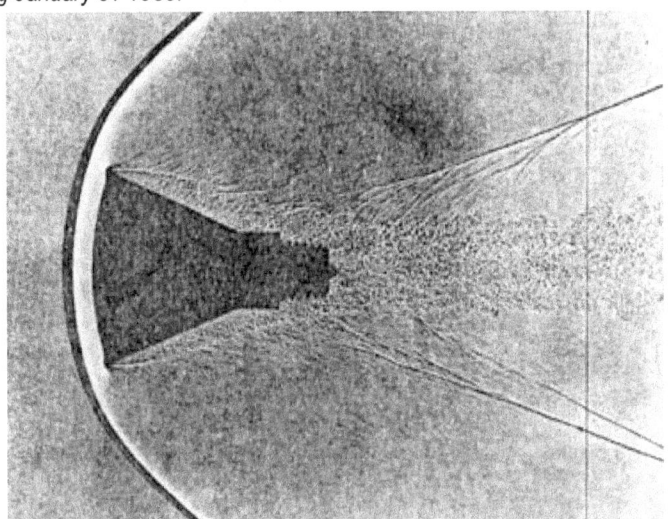

Figure 6. Shadowgraph of spacecraft model in Ames Supersonic Free-Flight Pressurized Range.

February 11

Space Task Group and Army Ballistic Missile Agency personnel met at Huntsville, Alabama, to discuss Redstone and Jupiter flight phases of Project Mercury. During the course of the meeting the following points became firm: (1) Space Task Group was the overall manager and technical director of this phase of the program, (2) ABMA was responsible for the launch vehicle until spacecraft separation, (3) ABMA was responsible for the Redstone launch vehicle recovery (this phase of the program was later eliminated since benefits from recovering the launch vehicle would have been insignificant), (4) Space Task Group was responsible for the spacecraft flight after separation, (5) McDonnell was responsible for the adapters for the Mercury-Redstone configuration, and (6) ABMA would build adapters for the Mercury-Jupiter configuration. Because many points could only be settled by detailed design studies, it was decided to establish several working panels for later meetings.

> Memo, Paul E. Purser, Space Task Group, to Project Mercury Director, subject: Project Mercury Meeting on February 11, 1959, at ABMA, February 17, 1959. Memo, George Low to NASA Administrator, subject: Status Report No. 8, Project Mercury, Mar. 4, 1959.

February 12

Search and recovery support by the Navy was discussed in a meeting with officials of that service and NASA. At the end of the conference, a NASA-Navy Committee was formed to work out a detailed plan. NASA members included E. C. Buckley, C. W. Mathews, and G. M. Low. The Navy was represented by Captain J. W. Gannon, with other members to be chosen at a later date.

> Memo, George Low to NASA Administrator, subject: Status Report No. 7, Project Mercury, Feb. 17, 1959.

February 12-13

Discussions were held at Langley Field between the Space Task Group and the Air Force Ballistic Missile Division covering aspects of the use of Atlas launch vehicles in Project Mercury. Specifically discussed were technical details of the first Atlas test flight (Big Joe), the abort sensing capability for later flights, and overall program objectives.

> Memo, A. C. Bond to Director of Project Mercury, subject: Visit of Ballistic Missile Division, Space Technology Laboratories, and Convair Representatives to Space Task Group on February 12 and 13, 1959, regarding Atlas Booster for Project Mercury, Feb. 18, 1959.

February 15

The medical examinations at the Wright Air Development Center for the final selection of the Mercury astronauts were started.

> Memo, George Low to NASA Administrator, subject: Status Report No. 8, Project Mercury, Mar. 4, 1959.

February 17

The first formal meeting of the Navy-NASA Committee on Project Mercury search and recovery operations was held. They decided that joint recovery exercises would be initiated as soon as possible. The committee members determined that the Navy, particularly the Atlantic fleet, could support operations from Wallops Island; could perform search and recovery operations along the Atlantic Missile Range, using of the selected Project Mercury vehicles; and that naval units could support operations in the escape area between Cape Canaveral and Bermuda.

> Memo, George Low to NASA Administrator, subject: Status Report No. 8, Project Mercury, Mar. 4, 1959.

Members of the Space Task Group, Langley, Ames, McDonnell and NASA Headquarters drafted a coordinated program for wind tunnel and free-flight tests in support of Project Mercury.

> Memo, Abe Silverstein, Director of Space Flight Development, NASA Hq., to Director of Aeronautical and Space Research, NASA Hq., subject: Langley and Ames Research Center Support for Project Mercury, March 6, 1959, and two inclosures.

February 19

In a speech, Dr. T. K. Glennan estimated that Project Mercury would cost over $200 million. The cost, he said in effect, was high because a new area of technology was being explored for the first time and there were no precedents or experience factors from which to draw, and because the world-wide tracking network construction was a tremendous undertaking.

> Draft Memorandum, John H. Disher to David Williamson, NASA Hq., no subject, March 31, 1960.

February 20

Responsibility for planning and contracting for Project Mercury tracking facilities was formally assigned to the Langley Research Center (see January 16, 1959, entry).

> Memo, NASA Hq., to Langley, subject: Request the Langley Research Center Assume Responsibilities for Project Mercury Instrumentation Facilities, Feb. 20, 1959.

February 24

Mercury-Redstone-Jupiter Study Panel Number IV (choice of trajectory, aerodynamics, and flight loads) met at Redstone Arsenal. Subjects studied included pilot safety, simulation of entry from orbit, length of zero-g time, missile stability and aerodynamics, ascent accelerations, and range. This group reconvened on March 13, 1959.

> Report No. 1, Mercury-Redstone-Jupiter Study Panel No. IV, March 20, 1959.

February 26

Panel Number I (Design Subcommittee) met at Redstone Arsenal for the first time to discuss integration requirements for the Mercury spacecraft with the Redstone and Jupiter launch vehicles.

> Memo, William M. Bland, Jr., to Director of Project Mercury, subject: First Meeting of Panel Number 1 Held February 26, 1959, at ABMA, Huntsville, Alabama, March 4, 1959.

Space Task Group and Langley Research Center personnel visited the Arnold Engineering Development Center, Tullahoma, Tennessee, to ascertain if the AEDC facilities were equipped to perform tests on scale models of the Mercury spacecraft and to arrange a testing schedule.

> Memo, Albin O. Pearson to Associate Director, NASA Langley, subject: Visit of NASA Personnel to AEDC, Tullahoma, Tenn., for the Purpose of Discussing the Testing of Models of the McDonnell (Project Mercury) Capsule in the AEDC Facilities, March 5, 1959.

February 27

Space Task Group personnel established the design trajectory for the Big Joe flight test. Convair Astronautics and Space Technology Laboratories personnel provided consultation and advice on ways in which these trajectory requirements could be met.

> Memo, Christopher C. Kraft to Director, Project Mercury, subject: Meeting with Space Technology Laboratories and Convair Representatives on Feb. 27, 1959, to Discuss Design Trajectories for First Atlas-Capsule Ablation Test, March 2, 1959.

February (during the month)

During a meeting between personnel of the Space Task Group and the Air Force Ballistic Missile Division, the responsibilities of the two organizations were outlined for the first two Atlas firings. Space Technology Laboratories, under Air Force Ballistic Missile Division direction, would select the design trajectories according to the specifications set forth by the Space Task Group. These specifications were to match a point in the trajectory at about 450,000 feet, corresponding to a normal reentry condition for the manned spacecraft after firing of the retrorockets at an altitude of 120 nautical miles. Space Technology Laboratories would also provide impact dispersion data, data for range safety purposes, and the necessary reprogramming of the guidance computers. The spacecraft for the suborbital Atlas flights would be manufactured under the direction of the Lewis Research Center, based on Space Task Group designs. Space Task Group was developing the spacecraft instrumentation, with a contingent of personnel at the Lewis Research Center. The attitude control system was being developed by Lewis.

> Memo, George Low to NASA Administrator, subject: Status Report No. 8, Project Mercury, Mar. 4, 1959.

Six working panels concerned with various aspects of the Mercury-Redstone program were formed to resolve problem areas that might arise. Later the number of panels was reduced to four, and then to three. Typical areas of study included design coordination, pilot safety, and aerodynamics, to name a few.

> NASA Space Task Group, *Project Mercury [Quarterly] Status Report No. 5 for Period Ending January 31, 1960.*

March 6

Space Task Group and McDonnell officials met in St. Louis, Missouri, to discuss spare part and ground support equipment requirements for Project Mercury.

Shortly thereafter, McDonnell submitted a preliminary plan for spare parts and check-out equipment to Space Task Group and NASA Headquarters for review.
>Memo, George Low, Chief, Manned Space Flight Development, NASA Hq., subject: Proposed Contract Amendments, Project Mercury Capsule, March 12, 1959.

March 8

An abort test was conducted at Wallops Island on a full-scale model of the spacecraft with the escape tower, using a Recruit escape rocket. The configuration did not perform as expected (erratic motion), and as a result, the Langley Research Center was requested to test small-scale flight models of the abort system to determine its motion in flight.
>Memo, Howard S. Carter and Carl A. Sandahl to Associate Director, NASA-Langley, subject: Weekly Progress Report for Week of March 8, 1959, on Langley Support of Project Mercury, March 16, 1959.

March 9

The Langley Research Center began exploratory noise transmission tests. The Center had also completed a report on rocket engine noise for use in determining the level of noise to which the prototype Mercury spacecraft would be subjected.
>Memo, Howard S. Carter and Carl A. Sandahl to Associate Director, NASA-Langley, subject: Weekly Progress Report for Week of March 8, 1959, on Langley Support of Project Mercury, March 16, 1959.

Tests were in progress at Langley and Wallops Island on several types of ablating materials under environmental conditions that would be experienced by a spacecraft reentering from orbit.

>Memo, Howard S. Carter and Carl A. Sandahl to Associate Director, NASA-Langley, subject: Weekly Progress Report for Week of March 8, 1959, on Langley Support of Project Mercury, March 16, 1959.

March 10

The Space Task Group was notified by McDonnell that several of its subcontractors were experiencing difficulties in procuring material necessary to fabricate Project Mercury components. This delay was being caused by the lack of a DX priority procurement rating.
>Letter, McDonnell Aircraft Corporation to NASA-Langley, subject: NAS 5-59, Effect of DO Priority Rating on Delivery Schedule, March 10, 1959.

March 11

Langley's Pilotless Aircraft Research Division conducted, at Wallops Island, the first full-scale test simulating a pad-abort situation. A full weight and size spacecraft was used. For the first 50 feet the flight was essentially straight, indicating the successful functioning of the abort rocket. Thereafter, the

spacecraft pitched through several turns and impacted a short distance from the shore. The malfunction was traced to the loss of a graphite insert from one of the three abort rocket nozzles, which caused a misalignment of thrust.
>Memo, Howard S. Carter to Associate Director, NASA-Langley, subject: Progress Report for Week of March 15, 1959, on Langley Support of Project Mercury, March 25, 1959; Memo, George Low to NASA Administrator, subject: Status Report No. 10, Project Mercury, March 24, 1959.

March 16

Purchase approval in the amount of $125,000 was requested by the Space Task Group from NASA Headquarters for the procurement of five developmental pressure suits for Project Mercury.
>Message, NASA 169, NASA-Langley to NASA Hq., March 16, 1959.

March 17

Funds were requested to purchase 6 main parachute and 12 drogue parachute canisters (fig. 7) from the Goodyear Aircraft Corporation in support of the Little Joe and Big Joe phases of Project Mercury.
>Memo, Andre J. Meyer, Jr., Space Task Group Chief, Engineering and Contract Administration Division, subject: Contracting of Parachute Canisters for Little Joe and Big Joe Development Launchings for Project Mercury, March 20, 1959.

Figure 7. Equipment installation in the parachute canister.

March 17-18

A Mock-Up Inspection Board meeting was held at the McDonnell plant to review the completed spacecraft mock-up. (See fig. 8.) As a result of this meeting, the contractor was directed to restudy provisions made for pilot egress; rearrange crew space to make handles, actuators, and other instruments more accessible to the pilot; and modify the clock, sequence lights, and other displays. This same type of meeting was held on many subsequent occasions to review production spacecraft.

Project Mercury Model 133 Mock-Up Review, Rpt. No. 6727, McDonnell Aircraft Corporation, March 17-18, 1959.

Figure 8. McDonnell mock-up of Mercury spacecraft including Atlas adapter and escape system.

March 20

John H. Disher was appointed as coordinator of the study panels. The purpose of this function was to prepare a unified source of information for organizations involved in the Mercury Program. The objective was to bring program plans and proposals together at a central location.
> Memo, Robert R. Gilruth to Langley Space Task Group, subject: Coordination of Meetings of Study Panels for Mercury Capsule Booster Systems, March 20, 1959.

Mercury-Redstone and Mercury-Jupiter test objectives were discussed in a meeting at Langley between Space Task Group and Army Ballistic Missile Agency personnel. At that time it was decided that the first flights of both the Redstone and Jupiter would be unmanned. The second flights would be "manned" with primates, and the Jupiter phase would end at that point. The six remaining Redstones would be used in manned flights for astronaut training.

> Memo, Walter J. Kapryan, subject: Project Mercury Meeting on March 20, 1959 at Langley Field, Virginia, March 26, 1959.

Space Task Group personnel prepared a study on the "Recovery Operations for Project Mercury" covering plans for suborbital and orbital flights. This document was forwarded to the Department of Defense for comment and for briefing of appropriate units.

> NASA-Space Task Group Study, *Recovery Operations for Project Mercury*, March 20, 1959.

March 23

As of this date, the McDonnell Aircraft Corporation listed some 32 items that required a DX priority procurement rating in support of Project Mercury. This highest national priority procurement rating had been requested by NASA on November 14, 1958.
> Letter, McDonnell Aircraft Corporation to George Low, NASA Hq., subject: NAS 5-59, Items Which Require DX Priority Rating, March 23, 1959.

March 26

The Langley Research Center received approval for funds to conduct hypersonic flight tests for the Mercury spacecraft. Langley's Pilotless Aircraft Research Division would conduct tests on heat transfer rates at a velocity of mach 17, and dynamic behavior tests from a velocity of mach 10 to a subsonic speed.
> Memo, NASA Director of Aeronautical and Space Research to Director, Space Flight Development, subject: Transfer of Funds to Langley Research Center for PARD Flight Testing of Project Mercury Capsule, March 20, 1959, and March 26, 1959, approval.

Space Task Group, Langley Research Center, and Air Force School of Aviation Medicine personnel met to plan bio-pack experiments that would be placed in several of the Little Joe research and development test flights.

> Minutes of Meeting, Project Mercury, subject: Bio-Paks for Little Joe Flights 2, 3, and 4, June 18, 1959.

March 27

Dr. T. Keith Glennan, the NASA Administrator, provided instructions for the marking of vehicles launched for the NASA, including the Mercury spacecraft. He stated that policy would be to paint UNITED STATES in bold block form.
> Memo, Floyd L. Thompson, Acting Director, Langley Research Center, subject: Identification of Vehicles Launched for NASA, April 15, 1959.

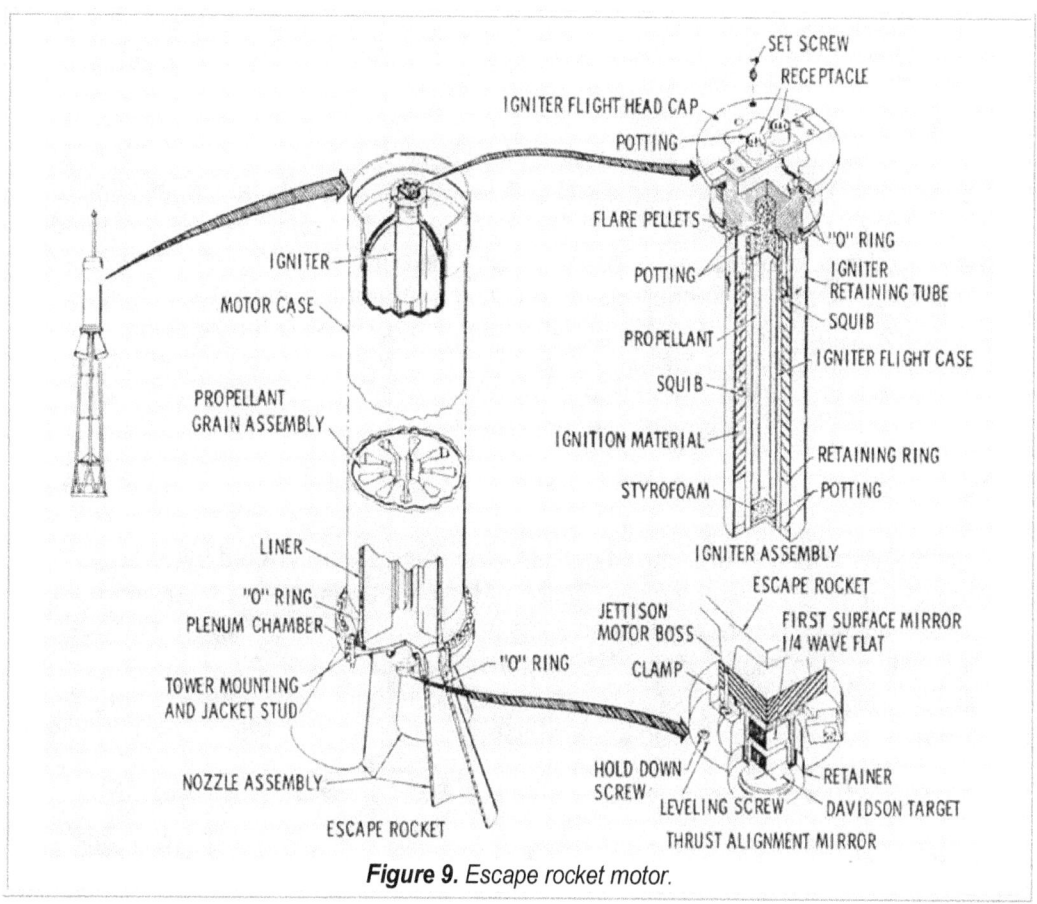

Figure 9. Escape rocket motor.

March 28

Space Task Group officials were involved in an investigation as to whether the escape system should be changed. In the original proposal, McDonnell's plan was to use eight small rockets housed in a fin adapter, but this plan was set aside for a NASA developed plan in which a single-motor tripod would be used. Later, during a test of the escape system, the escape rockets appeared to fire properly but the spacecraft began to tumble after launch. This tumbling action caused concern, and Space Task Group engineers felt that the tower-escape system might have to be discarded, and a "second look" was taken at the McDonnell proposal. The engineers concluded, however, that there were too many problems involved and the single-motor tripod concept was retained and has been proven to be quite effective. (See fig. 9.)

> Memo, Howard S. Carter to Associate Director, NASA-Langley, subject: Progress Report for Week of March 29, 1959, on Langley Support of Project Mercury, April 7, 1959.

March 29

Studies were in progress to determine the optimum altitude for separation of the Little Joe spacecraft from its launch vehicle.

Memo, Howard S. Carter to Associate Director, NASA-Langley, subject: Progress Report for Week of March 29, 1959, on Langley Support of Project Mercury, April 7, 1959.

March 30

Space Task Group personnel visited the Atlantic Missile Range at the invitation of the Army Ballistic Missile Agency to observe a Jupiter launch vehicle firing and the procedures followed on the day preceding the firing. The group toured the blockhouse and received briefings on various recorders that might be used in the centralized control facility for Mercury-Redstone and Mercury-Jupiter flights.

Memo, Christopher C. Kraft, Jr., to Director of Project Mercury, subject: Visit to AFMTC on March 13, and March 30, 1959, to Witness a Jupiter Dry-Run Procedure and Talk with AFMTC Range Safety Personnel, April 13, 1959.

March 31

Range Safety personnel at the Atlantic Missile Range were briefed by Space Task Group personnel on the description of the Mercury spacecraft, how it would function during a normal flight on an Atlas launch vehicle, and suggest methods for initiation of an abort during different powered phases of a flight. Atlantic Missile Range personnel discussed their past experience, and work was started to draft a Project Mercury range safety plan.

Memo, William M. Bland, Jr., and Christopher C. Kraft, Jr., to Director of Project Mercury, subject: Meeting with Range Safety People at AFMTC, March 31, 1959, April 3, 1959.

April 2

A preliminary briefing was conducted for prospective bidders on construction of the worldwide tracking range for Project Mercury. This meeting was attended by representatives from 20 companies. At this time the preliminary plan called for an orbital mission tracking network of 14 sites. Contacts had not been made with the governments of any of the proposed locations with the exception of Bermuda. It was planned that all the sites would have facilities for telemetry, voice communications with the pilot, and teletype (wire or radio) communications with centers in the United States for primary tracking. The tracking sites would provide the control center at Cape Canaveral, Florida, with trajectory predictions; landing-area predictions; and vehicle, systems, and pilot conditions.

Memo, George Low to NASA Administrator, subject: Status Report No. 11, Project Mercury, April 6, 1959; Project Mercury [Quarterly] Status Report No. 2 for Period Ending April 30, 1959.

Crew selection for Project Mercury was completed, resulting in the selection of seven astronauts to participate in the Mercury program.

Memo, George Low to NASA Administrator, subject: Status Report Project Mercury, April 6, 1959.

April 2-16

NASA and the military services conducted meetings to draft final plans for the Project Mercury animal payload program. The animal program was planned to cover nine flights, involving Little Joe, Redstone, Jupiter, and Atlas launch vehicles.
> Memo, George Low to NASA Administrator, subject: Status Report No. 12, Project Mercury, April 16, 1959.

An initial orientation was given to the seven Project Mercury astronauts, when they reported to the Space Task Group for duty.

> Memo, Warren J. North to NASA Administrator, subject: Background of Project Mercury Schedules, Aug. 14, 1960.

After responsibility for the worldwide tracking range construction of Project Mercury had been assumed by the Langley Research Center, the following study contracts were placed: (1) Aeronutronics to study radar coverage and trajectory computation requirements, (2) RCA Service Corporation for specification writing, (3) Lincoln Laboratories for consultant services and proposal evaluations, and (4) Space Electronics for the design of the control center at Cape Canaveral.

> Memo, George Low to NASA Administrator, subject: Status Report No. 11, Project Mercury, April 6, 1959.

The Chief of Naval Operations directed the Atlantic Fleet to support Project Mercury as follows: (1) landing and recovery systems in the area of Norfolk, Virginia, to develop spacecraft pickup and handling techniques for ships and helicopters, (2) recovery of capsules on solid rocket launch vehicle tests in the area of Wallops Island, and (3) Atlas launch vehicle development or ballistic flights from the Atlantic Missile Range. Details for orbital flight support had not been accomplished at that time.

> Memo, George Low to NASA Administrator, subject: Status Report No. 11, Project Mercury, April 6, 1959.

April 9

At a press conference in Washington, D.C., Dr. T. Keith Glennan announced that seven pilots had been selected for the Mercury program. These were Lt. Commander Alan B. Shepard, Jr., Navy; Captain Virgil I. Grissom, Air Force; Lt. Colonel John H. Glenn, Jr., Marines; Lieutenant Malcolm Scott Carpenter, Navy; Lt. Commander Walter M. Schirra, Jr., Navy; Captain Donald K. Slayton, Air Force; Captain Leroy Gordon Cooper, Jr., Air Force. (See fig. 10.)
> Hearing before the Committee on Science and Astronautics, U.S. House of Representatives, 86th Congress, 1st Session, Meeting with the Astronauts, Project Mercury, Man-in-Space Program, May 28, 1959.

Figure 10. The seven Mercury astronauts: L to R: Carpenter, Cooper, Glenn, Grissom, Schirra, Shepard and Slayton.

Figure 11. Scale model of escape tower configuration tested at Arnold Engineering Development Center.

April 9-10

Investigations of two escape configurations for Mercury spacecraft were conducted in a 16-foot transonic circuit at the Arnold Engineering Development Center, Tullahoma, Tennessee, for determination of static stability and drag characteristics of the configurations. (See fig. 11.)
"A Chronology of the Arnold Engineering Development Center"; *History of Arnold Engineering Development Center*, January-June 1959, Vol. I, pp. 38-41.

April 10

Escape-motor canting-angle tests were completed at Wallops Island. Tests were conducted in 5 degree increments between 10 degrees to 30 degrees, and visually it appeared stability was better at the larger angle.

> Memo, Howard S. Carter to Associate Director, NASA Langley, subject: Progress for Week of April 12, 1959, on Langley Support of Project Mercury, April 21, 1959.

April 12

Tests were in progress at Langley in which an aluminium honeycomb structure was used partially to absorb the spacecraft impact load. (See fig. 12.) Robert R. Gilruth, Project Mercury Director, had stated his belief of this requirement on January 16, 1959.

> Memo, Howard S. Carter to Associate Director, NASA Langley, subject: Progress for Week of April 12, 1959, on Langley Support of Project Mercury, April 21, 1959.

Space Task Group conducted the second full-scale beach abort test on Wallops Island. A deliberate thrust misalignment of 1 inch was programed into the escape combination. Lift-off was effected cleanly, and a slow pitch started during the burning of the escape rocket motor. The tower separated as scheduled and the drogue and main parachutes deployed as planned. The test was fully successful.

> Memo, Howard S. Carter to Associate Director, NASA Langley, subject: Progress for Week of April 12, 1959, on Langley Support of Project Mercury, April 21, 1959.

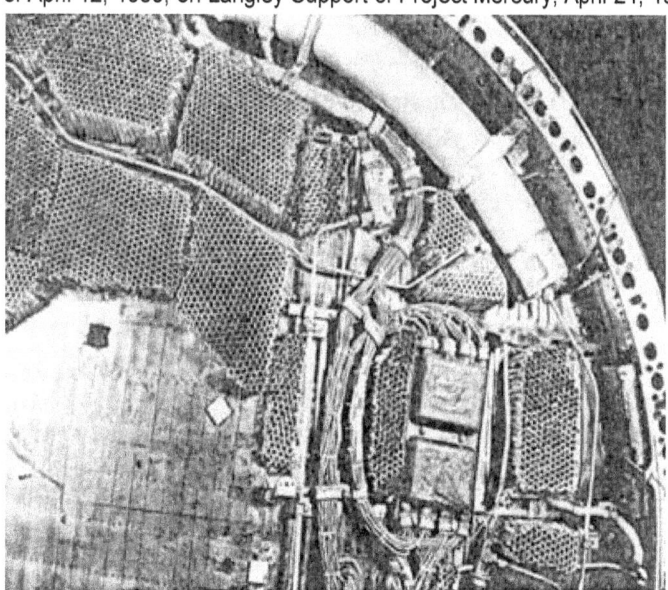

Figure 12. Honeycomb structure partially to absorb impact force.

April 13

Two small-scale spacecraft escape-tower combinations were launched successfully at Wallops Island, and on the next day a full-scale spacecraft escape system was launched. The complete sequence of events - escape system firing, escape tower jettisoning, parachute deployment, landing, and helicopter recovery - was satisfactory.

> Memo, George Low to NASA Administrator, subject: Status Report No. 12, Project Mercury, April 16, 1959.

NASA placed a request with the Navy for the use of its Aviation Medical Acceleration Laboratory at Johnsville, Pennsylvania. NASA desired to use the laboratory's AMAL human centrifuge in support of the Mercury astronaut training program.

> Letter, Warren J. North, NASA, to Captain F. K. Smith, Director, AMAL, subject: Request for use of Centrifuge at AMAL, Johnsville, April 13, 1959.

Rear Admiral J. W. Gannon was appointed by Donald A. Quarles, Deputy Secretary of Defense, to head a Department of Defense group to study with NASA the recovery aspects of Project Mercury.

> Letter, Deputy Secretary of Defense to Dr. Glennan, no subject, April 13, 1959.

April 15

Ground-instrumentation requirements for firing Little Joe test vehicles at Wallops Island were drafted. These requirements involved pulse radars, camera, Doppler radar, wind-monitoring instruments, telemetry equipment, and a ground destruct system.

> Memo, Charles H. McFall to Associate Director, NASA Langley, subject: Project Little Joe: Ground Instrumentation Required During Firing of Little Joe models at Wallops Island, April 15, 1959.

April 16

NASA and the military services held a meeting to discuss the search and recovery aspects of Project Mercury. Admiral Gannon, the service spokesman, stated that the meeting was exploratory but that the Navy and other services would support the project.

> Memo, George Low to NASA Administrator, subject: Status Report No. 12, Project Mercury, April 16, 1959.

Space Task Group, Langley Research Center, and Lewis Research Center personnel met to discuss development plans regarding construction and instrumentation of Big Joe Number I reentry spacecraft test vehicle. During the course of this meeting, milestone objectives of the work to be accomplished were drafted.

Memo, Howard S. Carter to Associate Director, NASA Langley, subject: Progress for Week of April 19, 1959, on Langley Support of Project Mercury, April 27, 1959.

NASA requested that the Air Force furnish two TF-102B and two T-33 aircraft to be used by the Project Mercury astronauts. One of the requirements in the astronaut training program was to maintain proficiency in high performance aircraft.

Memo, George Low to NASA Administrator, subject: Status Report No. 12, Project Mercury, April 16, 1959.

April 22

In a meeting at Langley, NASA officials concluded that the tower configuration was the best escape system for the Mercury spacecraft and development would proceed using this concept. (See fig. 13.) However, limited studies of alternate configurations would continue (see March 28, 1959, entry).

Memo, George Low to NASA Administrator, subject: Status Report No. 13, Project Mercury, May 6, 1959.

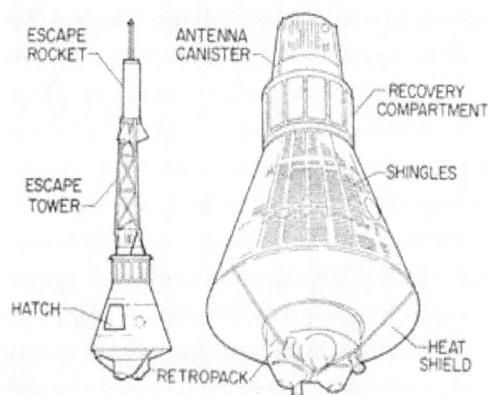

Figure 13. Spacecraft and escape system configuration.

April 23-24

A meeting was held at Langley to coordinate the activities of individuals who would be engaged in handling, reducing, and analyzing data received from the Big Joe spacecraft. Procedures for data pickup and for supplying the information to the appropriate installation were established. A majority of the data reduction workload was carried out by the Lewis Research Center and the Space Task Group.

Memo, M. J. Krasnican to Space Task Group records, subject: Coordination Meeting on Data Handling, Reduction, and Analysis for Big Joe Capsule held at Langley Space Task Group, April 23, 24, 1959, Apr. 28, 1959.

April 27

Project Mercury was accorded the DX priority procurement rating.
> Memo, George Low to NASA Administrator, subject: Status Report No. 13, Project Mercury, May 6, 1959.

The seven Project Mercury astronauts reported for duty and their training program was undertaken immediately.
> NASA Space Task Group, *Project Mercury [Quarterly] Status Report No. 2 for Period Ending April 30, 1959*; Memo, George Low to NASA Administrator, subject: Status Report No. 13, Project Mercury, May 6, 1959.

A tentative schedule of astronaut activities for the first months of training was issued. Actual training began the next day. Within 3 months the astronauts were acquainted with the various facets of the Mercury program. The first training week was as follows: Monday, April 27, check in; April 28, general briefing; April 29, spacecraft configuration and escape methods; April 30, support and restraint; May 1, operational concepts and procedures. These lectures were presented by specialists in the particular field of study. Besides the above, unscheduled activities involved 3 hours flying time and 4 hours of athletics.
> Tentative Schedule of Activities for First Months of Training Program (beginning Monday, April 27, 1959).

April 27-28

The Department of Defense working group on Mercury search and recovery operations met at Patrick Air Force Base, Florida, to establish service responsibilities and support for the first two Mercury-Atlas ballistic flights.
> Memo, George Low to NASA Administrator, subject: Status Report No. 13, Project Mercury, May 6, 1959.

April (during the month)

In the recovery landing system, the extended-skirt main parachute was found to be unsafe for operation at altitudes of 10,000 feet and was replaced by a "ring-sail" parachute of similar size. This decision was made after a drop when the main parachute failed to open and assumed a "squidding" condition. Although little damage was sustained by the spacecraft on water impact, parachute experts decided that the ring-sail configuration should be adopted, and the air drop spacecraft were fitted.
> NASA Space Task Group, *Project Mercury [Quarterly] Status Report No. 2 for Period Ending April 30, 1959*; Memorandum, Robert R. Gilruth, Director of Project Mercury, to NASA Hq., subject: Required Basic Research on Parachutes to Support Manned Space Flight, July 6, 1959.

May 1

A Little Joe Project Coordination Meeting, attended by personnel from Space Task Group, McDonnell, and Wallops Island, was held for the first time. The purpose of the meeting was to determine the status of various developmental phases and whether or not proper coordination was being effected with other related projects in the Mercury program (Big Joe, Mercury-Atlas, Mercury-Redstone, and Mercury-Jupiter). The important factor with regard to the latter item was whether or not a reasonable launch schedule could be established and maintained.
> Memo, Ronald Kolenkiewicz and John B. Lee to Director, Project Mercury, subject: Coordination Meeting for Little Joe Project, May 6, 1959.

May 5

Space Task Group personnel held a meeting to discuss the complete recovery test program. Items of consideration included the availability of model spacecraft for the test, deciding the areas in which the tests would be held (Phase I - Wallops Island drops, and Phase II - Atlantic drops), and establishing the time schedule for the test program.
> Memo, Robert R. Gilruth, Director of Project Mercury, to Hartley A. Soule, Langley Research Center, subject: Request for Assistance of Langley Research Center in Project Mercury Capsule Drop Test Program, June 2, 1959, with inclosures.

May 6

Pigs were eliminated as Little Joe flight test subjects when studies disclosed that they could not survive long periods of time on their backs. However, McDonnell did use a pig, "Gentle Bess," to test the impact crushable support, and the test was successful.
> Memo, George Low to NASA Administrator, subject: Status Report No. 13, Project Mercury, May 6, 1959.

May 11

A spacecraft recovery study contract was awarded to Grumman Aircraft Corporation.
> Memo, Warren J. North to NASA Administrator, subject: Background of Project Mercury Schedules, Aug. 14, 1960.

A NASA policy concerning Mercury astronauts was issued. The astronauts were subject to the regulations and directives of NASA, and information of unclassified nature reported by the astronauts would be disseminated to the public. These were but two examples in the policy statement.

> Senate Report, No. 1014, 86th Congress, 1st Session, subject: *Project Mercury Man-in-Space Program of the National Aeronautics and Space Administration*, 86th Congress (Dec. 1, 1959).

May 12-14

An informal meeting of the Mock-Up Inspection Board was held at McDonnell to review changes to the spacecraft development program resulting from the March mock-up meeting. Besides the review, a number of suggestions were made for changes in the crew space layout to permit more effective use of the controls, particularly when the astronaut was in the pressure suit in a full-pressurized condition. Among suggested changes were the shoulder harness release, the spacecraft compression and decompression handles, the ready switch, and the spacecraft squib switch. Test subjects also found that when in the fully pressurized suit none of the circuit breakers could be reached. McDonnell was directed to act on these problem areas.
    Minutes of Mock-Up Review held at McDonnell Aircraft Corporation, May 12-14, 1959.

May 17

The Langley Research Center was in the process of preparing a one-fourteenth scale model of the Mercury spacecraft for launch from Wallops Island on a five-stage rocket to a speed of mach 18.
    Memo, Howard S. Carter to Associate Director, NASA Langley, subject: Progress for May 17-31, 1959, on Langley Support of Project Mercury, June 3, 1959.

May 21

Langley Specification Number S-45, entitled "Specifications for Tracking and Ground Instrumentation System for Project Mercury," was issued. Proposals were received from seven contractor teams by June 22, 1959, and technical evaluations were started.
    Memo, Howard S. Carter to Associate Director, NASA Langley, subject: Progress for May 17-31, 1959, on Langley Support of Project Mercury, June 3, 1959; Memo, H. J. E. Reid, Director NASA Langley, to NASA Hq., subject: Further Plans for the Procurement and Ground Instrumentation Systems for Project Mercury, June 26, 1959.

May 22

The Space Task Group, in the process of negotiations with the Army Ordnance Missile Command on the cost of Redstone and Jupiter boosters in support of Project Mercury, received revised funding estimates for study covering Contract HS-44 (Redstone) and HS-54 (Jupiter).
    Memo, Paul E. Purser to Director, Project Mercury, subject: Analysis of AOMC Revised Funding Estimates for Redstone and Jupiter, HS-44 and HS-54, June 5, 1959.

The Project Mercury balloon flight test program was canceled. The Space Task Group officials determined that the spacecraft could be tested environmentally in the Lewis Research Center's altitude wind tunnel. This included correct temperature and altitude simulations to 80,000 feet. The pilot could exercise the attitude control system and retrorockets could be fired in the tunnel. Because an active contract did exist with the Air Force, it was decided the two balloon drop tests with unmanned boiler-plate spacecraft would be accomplished.

Memo, George Low to NASA Administrator, subject: Status Report No. 14, Project Mercury, May 22, 1959.

May 25

A meeting was held at Johnsville, Pennsylvania, to consider astronaut training programs on the centrifuge. (See fig. 14.) During this meeting, Space Task Group personnel reviewed a draft memorandum prepared by the Aviation Medical Acceleration Laboratory concerning the methods they felt should be used. Also, possible centrifuge training periods for the astronauts were discussed, and tentative dates were set for August 1959 and January 1960.

Memo, Euclid C. Holleman to Chief, Research Division, NASA High-Speed Flight Station, subject: Meeting to Consider Plans for the Aviation Medical Acceleration Laboratory for the Next Fiscal Year, May 25, 1959.

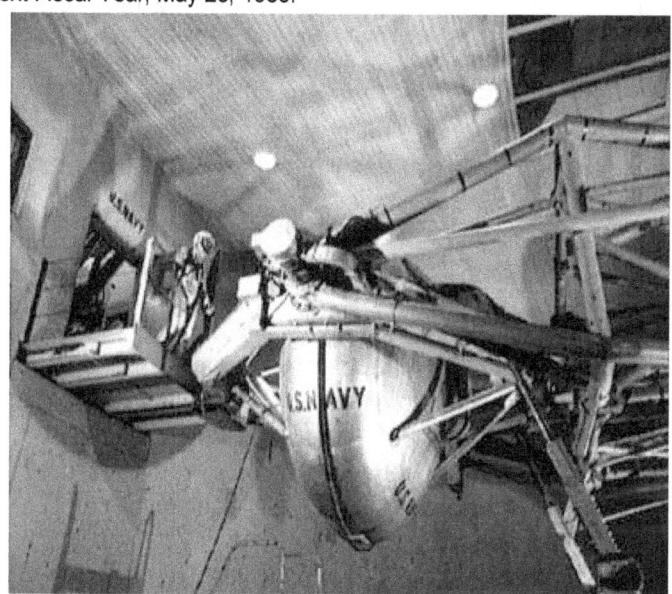

Figure 14. Human centrifuge at the Navy's Aviation Medical Acceleration Laboratory, Johnsville, Pennsylvania, used in Mercury astronaut training program.

May 28

North American Aviation delivered the first two Little Joe booster airframes, and noted that the four remaining were on fabrication schedule. The planned program was moving smoothly, for rocket motors to be used in the first flight were available at Wallops Station, Virginia, the test flight launching site. In addition, procurement of the test spacecraft incorporating Mercury flight items was on schedule, and the first spacecraft had been instrumented by Space Task Group personnel. Work was also in progress on other test spacecraft.

> Memo, Warren J. North to NASA Administrator, subject: Background of Project Mercury Schedules, August 14, 1960.

Primates Able and Baker, aboard an Army Jupiter missile nose cone, were launched 300 miles into space and landed 1,700 miles down range from the launch site at Cape Canaveral. Telemetry data disclosed that the responses of the animals were normal for the conditions they were experiencing. During the boost phase, when the higher g-loads were being sustained, body temperature, respiration, pulse rate, and heartbeat rose but were well within tolerable limits. During the weightless period along the trajectory arc, the physiological responses of Able and Baker approached normal - so near, in fact, that according to telemetry data, Baker appeared either to doze or to become drowsy. Upon reentry, the responses rose again, but at landing the animals were nearing a settled physiological state. This flight was another milestone proving that life could be sustained in a space environment.

> Grimwood, *History of the Jupiter Missile Program*, July 1962; House Committee Print No. 35, Hearing Before the Committee on Science and Astronautics, U.S. House of Representatives, Jupiter Missile Shot - Biomedical Experiments, June 3, 1959.

A quick-release, side exit hatch was designed for the spacecraft. The design consisted of a continuous double explosive train to assure that all bolts were actually broken upon activation of the device.

> NASA Space Task Group, *Project Mercury [Quarterly] Status Report No. 4 for Period Ending October 31, 1959.*

During this period, the astronauts and other NASA personnel devoted a great deal of study to the Mercury spacecraft cockpit. The following factors were under particular scrutiny: (1) routine and emergency flight procedures; (2) anthropometric dimensions of the seven astronauts, which had demonstrated flight safety inadequacies in the early layout of the cockpit; and (3) layout requirements which were reviewed according to the dimensions of the astronauts while wearing a full-pressure garment, in both routine unpressurized and pressurized states, and according to the astronaut's ability to reach any control under both routine and emergency conditions. (See fig. 15.)

> NASA Space Task Group, *Project Mercury [Quarterly] Status Report No. 4 for Period Ending October 31, 1959.*

June 1

Personnel strength in support of Project Mercury included 204 at the Space Task Group, 98 at the Langley Research Center, 44 at the Lewis Research Center, and 21 on the Mercury tracking network, for a grand total of 363.
> Chart, Space Task Group Complement Analysis, June 1, 1959.

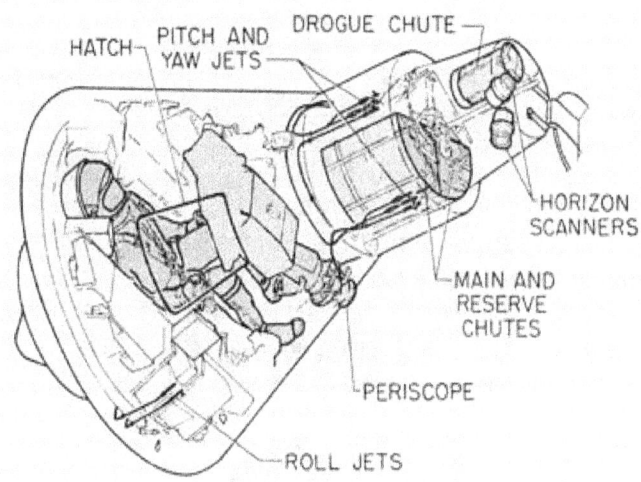
Figure 15. Spacecraft interior arrangement.

June 5

The drogue parachute configuration was changed from 19.5 percent porosity, flat circular ribbon chute to a 28 percent porosity, 30 degree conical canopy.

> NASA Space Task Group, *Project Mercury [Quarterly] Status Report No. 3 for Period Ending July 31, 1959.*

The Army Ballistic Missile Agency submitted a proposal (Report No. DG-TR-7-59) for a Mercury-Redstone inflight abort sensing system. This system would monitor performance of the control system (attitude and angular velocity), electrical power supply, and launch vehicle propulsion. If operational limits were exceeded, the spacecraft would be ejected from the launch vehicle and recovered by parachute.

> Study, Proposal for Mercury-Redstone Automatic Inflight Abort Sensing System by F. W. Brandner, prepared by the Army Ballistic Missile Agency, June 5, 1959.

Space Technology Laboratories and Convair completed an analysis of flight instrumentation necessary to support the Mercury-Atlas program. The primary objective of the study was to select a light-weight telemetry system. A system weighing 270 pounds was recommended, and the National Aeronautics and Space Administration concurred with the proposal.

> Letter, Space Technology Laboratories, Inc., to NASA Space Task Group, subject: Atlas Telemetry Configuration, Project Mercury Orbital Flights, June 5, 1959; Letter, NASA Hq. to E. B. Doll, Space Technology Laboratories, subject: Details of Atlas Telemetry System for Project Mercury Flights, July 7, 1959.

June 8

The Big Joe spacecraft for the reentry test was delivered to Cape Canaveral.
> Memo, Warren J. North to NASA Administrator, subject: Background of Project Mercury Schedules, Aug. 14, 1960.

The Space Task Group advised the Navy's Bureau of Aeronautics of Government-furnished survival items that McDonnell would package in containers. (See fig. 16.) These included desalter kits, dye marker, distress signal, signal mirrors, signal whistle, first aid kits, shark chaser, PK-2 raft, survival rations, matches, and a radio transceiver. Navy assistance was requested in the procurement of these items.
> Letter, NASA Space Task Group, Bureau of Aeronautics, Department of the Navy, subject: Project Mercury Survival Equipment, June 8, 1959.

Space Task Group officials met with representatives of the School of Aviation Medicine to discuss detailed aspects of the bio-packs to be used in the NASA Little Joe Flight program. The packs were to be furnished by the school. The purpose was to gather life support data that would be applicable to the manned flights of Project Mercury.
> Space Task Group Minutes of Meeting, subject: Bio-Packs for Little Joe Flights 2, 3, and 4, June 8, 1959, at Space Task Group, June 18, 1959.

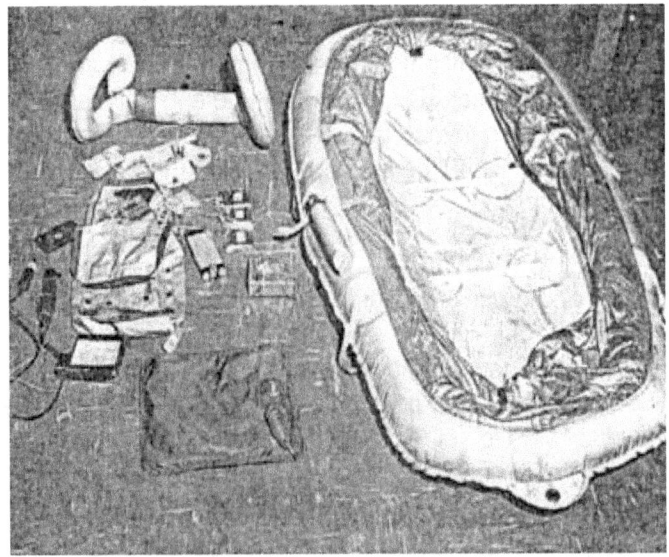

Figure 16. Astronaut survival equipment stowed in Mercury spacecraft.

June 12

A Source Selection Panel and a Technical Evaluation Board were organized and manned at the Langley Research Center to evaluate Mercury tracking and ground instrument action proposals. Technical evaluation of proposals was started on June 23, 1959, with seven companies under consideration. These were - in addition to Western Electric - Aeronutronics, Radio Corporation of America, Pan American Airways, Brown and Root, Chrysler Corporation, and Philco Corporation.

> Memo, H. J. E. Reid, Director of Langley Research Center to All Concerned, subject: Designation of Organization, Membership and Operating Procedures for the Source Selection Panel and the Technical Evaluation Board - Tracking and Ground Instrumentation, Project Mercury, June 12, 1959.

June 14-27

A visit was made to McDonnell and it was learned that the Mercury spacecraft was being designed structurally to withstand 149 decibels overall noise level. McDonnell, however, anticipated that the actual maximum level would not be above 128 decibels. Space Task Group personnel felt that even the 128 decibels were too high for pilot comfort, and extensive research toward the resolution of this matter was started.

> Memo, Howard S. Carter to Associate Director, NASA Langley, subject: Progress for June 14-27, 1959, on Langley's Support of Project Mercury, June 30, 1959.

June 18

A centrifuge program was conducted at Johnsville, Pennsylvania, to investigate the role of a pilot in the launch of a multi-stage vehicle. Test subjects were required to perform boost-control tasks, while being subjected to the proper boost-control accelerations. The highest g-force experienced was 15, and none of the test subjects felt they reached the limit of their control capability. As a note of interest, one of the test subjects, Neil Armstrong, was later selected for the Gemini program in September 1962.

> Memo to Chief, High-Speed Flight Station, subject: Summary of Boost Centrifuge Program.

June 19

The Mercury Capsule (spacecraft) Coordination Office was organized within the Space Task Group, with J. A. Chamberlin appointed head of the office. Duties were divided into four major categories as follows: (1) loads, thermodynamics, structures, and aerodynamics; (2) cabin, life support, and controls; (3) electronics, recovery, and sequencing; and (4) transportation and handling, schedules and testing, and standards and specifications. This action assured continuity of effort in monitoring the McDonnell contract. Also, this office arranged and coordinated meetings with McDonnell personnel and served as a clearing house for all NASA-McDonnell contracts. The committee, of course, received a majority of its data from technical sources within the formal Space Task Group organization.

> Memo, Robert R. Gilruth to Space Task Group Division, Branch, and Section Chiefs and Heads, subject: Capsule Coordination Office, June 19, 1959; Summary of the Method of

> Monitoring the McDonnell Capsule Contract, prepared by Space Task Group, July 10, 1959.

A Capsule Review Board was established to review, at regular intervals, action taken by the Capsule Coordination Office. Paul E. Purser was appointed chairman, with division heads, Coordination Office head, and Project and Assistant Project Directors serving as members.

> Memo, Robert R. Gilruth to Space Task Group Division, Branch, and Section Chiefs and Heads, subject: Organization of Capsule Coordination Office, June 19, 1959.

June 24

Against an original estimated cost of $15.5 million for eight Redstone launch vehicles in support of Project Mercury, the final negotiated figure was $20.1 million.

> Chart, Revised Funding HS-44, Project Mercury (Redstone) prepared by Control Office, Army Ballistic Missile Agency, June 24, 1959.

June 25

Navy surface vessels and aircraft were used in a recovery operation after an airdrop of a spacecraft off the coast from Jacksonville, Florida. The spacecraft was purposely dropped 40 miles away from the predicted impact point and 45 miles away from the nearest ship. Recovery was effected in 2 and one half hours.

> NASA Space Task Group, *Project Mercury [Quarterly] Status Report No. 3 for Period Ending July 31, 1959.*

June 28

Between the cited date and July 11, 1959, 12 heat-transfer tests were made in the Preflight Jet Test facility at Wallops Island on several ablation materials being considered for use on the spacecraft afterbody (not heat shield) for the Little Joe flights. Test conditions simulated those of actual Little Joe trajectories. Of the materials used, triester polymer and thermolag demonstrated the capability to protect the spacecraft against expected heat loads.

> Memo, Howard S. Carter to Associate Director, NASA Langley, subject: Progress for June 28-July 11, 1959, on Langley Support of Project Mercury, July 15, 1959.

June 29

A longitudinal static stability investigation was carried out for the Mercury manned orbital spacecraft model in the 16-foot transonic circuit at the Arnold Engineering Development Center.

> History of Arnold Engineering Development Center, January-June 1959, Vol. I, pp. 38-41.

June (during the month)

The Space Task Group furnished several boilerplate spacecraft to DesFlotFour (naval unit involved in Project Mercury recovery plans) for use in developing detailed recovery techniques. (See fig. 17.)

> Memo, Howard S. Carter to Associate Director, NASA Langley, subject: Progress for June 14-27, 1959, on Langley Support of Project Mercury, June 30, 1959.

McDonnell selected Northrop as the subcontractor to design and fabricate the landing system for Project Mercury. Northrop technology for landing and recovery systems dated back to 1943 when that company developed the first parachute recovery system for pilotless aircraft. For Project Mercury, Northrop developed the 63-foot ring-sail main parachute. (See fig. 18.)

> Material supplied by Jerome Ringer, Public Relations Department, Northrop Ventura, Jan. 1963.

Figure 17. Recovery test spacecraft showing recovery aids.

Figure 18. Main 63-foot ringsail parachute.

July 1

The order for Jupiter launch vehicles in support of Project Mercury was canceled because the same or better data could be obtained from Atlas flights.

> Memo, Abe Silverstein, Director of Space Flight Development, NASA Hq., to Langley Space Task Group, subject: Cancellation of Mercury-Jupiter Program, July 1, 1959.

July 1-2

A pressure suit compatibility evaluation in the Mercury spacecraft mock-up was performed in suits submitted by the David Clark Company, B. F. Goodrich Company, and International Latex Company. Four subjects participated in the tests.
> McDonnell Aircraft Corporation, subject: Project Mercury Engineering Status Report, June 1 to August 1, 1959.

July 6

As a result of a discussion between Maxime A. Faget, Space Task Group, and John E. Naugle, Space Science Division, NASA Headquarters, it was concluded that there were several important scientific experiments in the field of energetic particles research that could be performed by placing packets of emulsion within the Mercury spacecraft. Work was started to determine a suitable packet location, along with other details associated with conducting such experiments.
> Memo, John W. Townsend, Jr., Assistant Director, Space Science and Satellite Applications, NASA Hq., to Robert R. Gilruth, Director of Project Mercury, subject: Energetic Particle Research - Project Mercury, July 6, 1959.

Results of the technical and management evaluations of Mercury tracking network proposals were presented to the Langley Research Center Source Selection Board.

> NASA Space Task Group, *Project Mercury [Quarterly] Status Report No. 3 for Period Ending July 31, 1959.*

July 12

An agreement was made with the Air Force for Space Task Group to place microphone pickups on the skin of the Atlas launch vehicle as a part of the instrumentation to measure noise level during the Big Joe-Atlas launching. Distribution of the microphones was as follows: one inside the Mercury spacecraft, three externally about midway of the launch vehicle, and one on the Atlas skirt.
> Letter, Charles J. Donlan, Associate Director of Project Mercury, to R. W. Costin, Bostrom Research Laboratories (no subject) July 29, 1960, with inclosures.

July 13

Spacecraft horizon scanner qualification tests were started.
> McDonnell Aircraft Corporation, subject: Project Mercury Engineering Status Report, June 1 to August 1, 1959.

The Western Electric Company and associates were announced as winner of the competition for construction of the Mercury tracking network.

NASA Space Task Group, *Project Mercury [Quarterly] Status Report No. 3 for Period Ending July 31, 1959.*

July 20

Negotiations for construction of the Mercury tracking network were started with the Western Electric Company and their subcontractors (Bendix Aviation, International Business Machines, Bell Telephone Laboratories, and Burns and Roe), and a letter contract was signed on July 30, 1959, for the entire range. This included radar tracking; telemetry receiving, recording, and display; communications to both the spacecraft and surface stations; and the computing and control facilities.

NASA Space Task Group, *Project Mercury [Quarterly] Status Report No. 3 for Period Ending July 31, 1959.*

The Space Task Group forwarded Big Joe postflight requirements to Pan American personnel at the Atlantic Missile Range for use in preparing their documents concerning postflight handling of the Mercury special test spacecraft.

Letter, Robert R. Gilruth, Director of Project Mercury, to B. Porter Brown, NASA Atlas-Mercury Test Coordinator (no subject), July 20, 1959, with inclosures.

July 21

Alterations to Building "S" at Cape Canaveral for Project Mercury support were discussed in a meeting at Cape Canaveral. (See fig. 19.) A target date of December 1, 1959, was set for project completion. Therefore, this meant that Vanguard activities would have to be phased out of the building.

NASA Space Task Group, *Project Mercury [Quarterly] Status Report No. 3 for Period Ending July 31, 1959.*

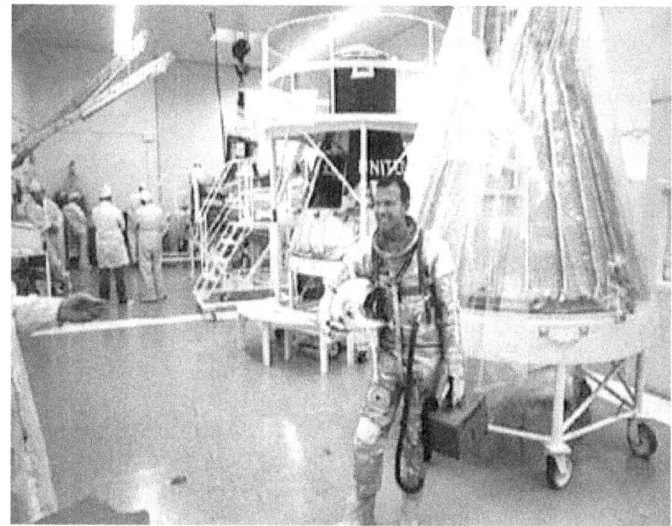

Figure 19. White room in Hanger S at Cape Canaveral.

July 22

The B. F. Goodrich Company was selected as the contractor to design and develop the Mercury astronaut pressure suit. Company technology in this field dated back to 1934, when it developed the first rubber stratosphere flying suit for attempts at setting altitude records.

> NASA Space Task Group, *Project Mercury [Quarterly] Status Report No. 3 for Period Ending July 31, 1959.*

A successful pad abort flight of a boilerplate spacecraft with a production version of the escape tower and rocket was made. (See fig. 20.) The escape rocket motor was manufactured by Grand Central Rocket, and the flight was the first operational test of this component.

> NASA Space Task Group, *Project Mercury [Quarterly] Status Report No. 3 for Period Ending July 31, 1959.*

The Space Task Group, McDonnell, and the Air Force Chart and Information Center held a meeting with regard to a map depicting Mercury spacecraft flight. At that time, it was decided that the chart would cover an area of 40 degrees latitude above and below the equator. The chart would show oceans and continents by colors to match probable visual characteristics. Orbit numbers and time since launch would be depicted and traced.

> NASA Space Task Group, *Project Mercury [Quarterly] Status Report No. 3 for Period Ending July 31, 1959.*

The Navy provided NASA with a list of reserve ships that might be used in direct support of Project Mercury, and on July 28, 1959, specific information was forwarded on ships that NASA might be interested in using.

> Letter, M. J. Luosey, Department of Navy, to NASA Langley Research Center (no subject), received July 28, 1959.

July 28

A boilerplate spacecraft, instrumented to measure sound pressure level and vibration, was launched in the second beach abort test leading to the Little Joe test series. The purpose of the instrumentation was to obtain measurement of the vibration and sound environment encountered on the capsule during the firing of the Grand Central abort rocket.

> Memo, Charles A. Hardesty to NASA Langley IRD files, subject: Sound Measurements on the Second Beach Abort Test on the Little Joe Capsule, Oct. 9, 1959.

July 30

Letter Contract NASA 1-430 was awarded to the Western Electric Company for construction of the Mercury tracking and ground instrumentation system. (See July 20, 1959 entry.)

    Memo, Sherwood L. Butler, Langley to NASA Headquarters, Code: BR, subject: Monthly Status Report - Project Mercury, Nov. 3, 1959.

Figure 20. Spacecraft with McDonnell designed escape system ready for firing at Wallops Island.

July 31

Personnel from the Aeromedical Field Laboratory inspected the first animal couch fabricated by McDonnell to be used in the Mercury animal flight program. The objective of the animal program was to provide verification of successful space flight prior to manned missions; to acquire data on physical and mental demands which will be encountered by the astronauts during space flight; to provide dynamic test of technical procedures and training for support personnel in handling the aeromedical program for manned flight; and to evaluate spacecraft environmental control systems and bioinstrumentation under flight conditions.

    NASA Space Task Group, *Project Mercury [Quarterly] Status Report No. 3 for Period Ending July 31, 1959.*

July (during the month)

The Mercury astronauts completed disorientation flights on the three-axis space simulator at the Lewis Research Center.

    McDonnell Aircraft Corporation, *Project Mercury Bi-Monthly Capsule Manufacturing and Tooling Report, August 1, 1959 to October 1, 1959*, p. 22.

Minneapolis-Honeywell delivered the first automatic stabilization and control system for the Mercury spacecraft to McDonnell. (See fig. 21.)

    NASA Space Task Group, *Project Mercury [Quarterly] Status Report No. 3 for Period Ending July 31, 1959.*

The Pilotless Aircraft Research Division of the Langley Research Center launched a 1/14th-scale model of the Mercury spacecraft at Wallops Island to a speed of Mach 3.5 and at an altitude of 40,000 feet. The model spacecraft went into a continuous tumble from separation to landing.

    Memo, Howard S. Carter to Associate Director, NASA Langley, subject: Progress for June 28-July 11, 1959, on Langley Support of Project Mercury, July 15, 1959.

Specialty assignments were made to each of the Mercury astronauts. Thus they became participating members in the NASA-McDonnell coordination meetings and the Mercury-Redstone or Mercury-Atlas meetings in their specialty area. Assignments were as follows: Scott Carpenter, navigation and navigational aids; Gordon Cooper, Redstone launch vehicle; John Glenn, crew space layout; Virgil Grissom, automatic and manual attitude control system; Walter Schirra, life-support system; Alan Shepard, tracking and recovery operations; and Donald Slayton, Atlas launch vehicle.

    NASA Space Task Group, *Project Mercury [Quarterly] Status Report No. 3 for Period Ending July 31, 1959.*

A three-axis hand controller and a pilot restraint system were delivered to NASA at the Johnsville centrifuge for use in the Mercury astronaut training program.

    McDonnell Aircraft Corporation, *Project Mercury Bi-Monthly Engineering Status Report, August 1, 1959 to October 1, 1959*, p. 22.

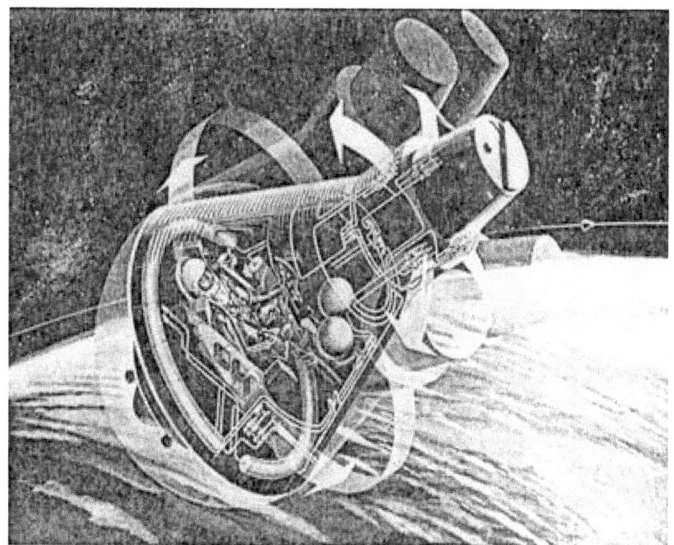

Figure 21. Spacecraft reaction control system.

August 3

Major General Donald N. Yates was appointed as the Department of Defense representative for Project Mercury support operations.
> Information supplied by Major General Leighton Davis' Office, April 1963.

August 4

Tests were started to check the operation of the redesigned Mercury drogue parachute. (See figs. 22 and 23.)
> NASA Space Task Group, *Project Mercury [Quarterly] Status Report No. 4 for Period Ending October 31, 1959.*

Figure 22. Vehicle for drogue parachute test at NASA Flight Research Center.

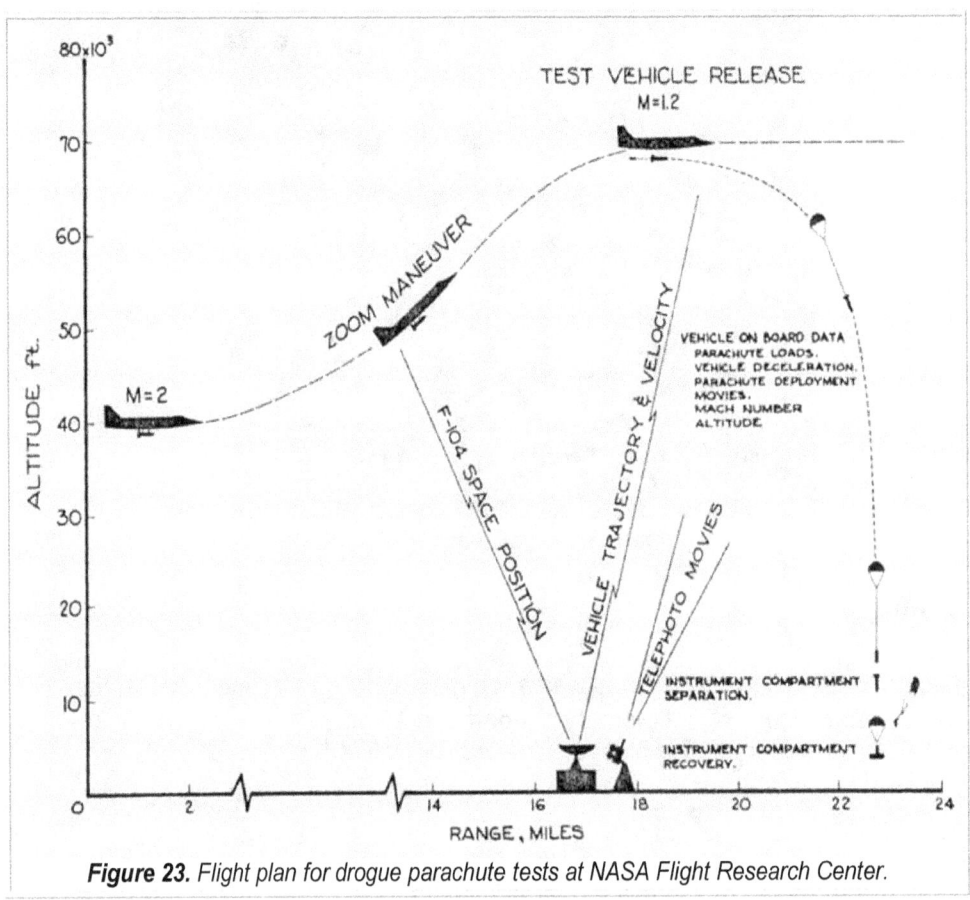

Figure 23. Flight plan for drogue parachute tests at NASA Flight Research Center.

August 6

Four F-102 aircraft were made available for use by the Mercury astronauts to maintain proficiency in high performance vehicles.
> Memo, Warren J. North to NASA Director of Space Flight Development, subject: Interim Status Report for Project Mercury, Aug. 7, 1959.

August 14

NASA Headquarters approved a Space Task Group proposal that negotiations be undertaken with McDonnell for the fabrication of six additional Mercury spacecraft.
> Memo, NASA Hq. to Langley Space Task Group, subject: Additional Capsules for Project Mercury, Sept. 9, 1959.

August 15

The astronauts began their initial centrifuge training at the Aviation Medical Acceleration Laboratory. During the first part of the month Space Task Group

personnel had installed and checked out Mercury spacecraft simulation equipment at the Aviation Medical Acceleration Laboratory in preparation for the astronaut centrifuge training program.
> Memo, Dr. W. S. Augerson, Life Systems Branch, to Chief, Flight Systems Division, Space Task Group, subject: Trip Report, Sept. 15, 1959.

August 21

During the countdown of the first programed Little Joe launching (LJ-1 beach abort test) at Wallops Island, the escape rocket fired prematurely 31 minutes before the scheduled launch. The spacecraft rose to an altitude of 2,000 feet and landed about 2,000 feet from the launch site. Premature firing was caused by a faulty escape circuit.
> NASA Space Task Group, *Project Mercury [Quarterly] Status Report No. 4 for Period Ending October 31, 1959.*

August 25

Testing was completed to check the effectiveness of the drogue parachute as a stabilizing device. The drogue parachute was fully qualified for deployment at speeds up to Mach 1.5 and altitudes of up to 70,000 feet. Ordinarily, during the operational phase of Project Mercury the drogue parachute was deployed at 40,000 feet, so the component well met operational requirements.
> NASA Space Task Group, *Project Mercury [Quarterly] Status Report No. 4 for Period Ending October 31, 1959.*

August 28

NASA Headquarters authorized the Space Task Group to enter into negotiations with the Air Force Ballistic Missile Division for the procurement of additional Atlas launch vehicles in support of Project Mercury. The authorization was to be incorporated into Contract No. HS-36.
> Memo, Warren J. North, Chief, Manned Satellite to Director, Space Flight Development, subject: Purchase Approval for Four Mercury Atlas Boosters, October 13, 1959.

August (during the month)

Qualification tests, which were started in May 1959, were completed for the 63-foot ringsail, main parachute. After this, complete parachute landing tests were initiated by spacecraft drops from a C-130 at Salton Sea, California.
> NASA Space Task Group, *Project Mercury [Quarterly] Status Report No. 4 for Period Ending October 31, 1959.*

McDonnell submitted its first monthly reliability report. The purpose of this report was to summarize the reliability efforts of McDonnell and its subcontractors in the design and development of the Mercury spacecraft.

NASA Space Task Group, *Project Mercury [Quarterly] Status Report No. 3 for Period Ending July 31, 1959.*

September 1-7

McDonnell moved a segment of its Mercury effort to Cape Canaveral in preparation for the operational phase of the program. Personnel were immediately assigned to committees to develop the plans for Mercury-Redstone and Mercury-Atlas missions. The McDonnell office was located in Hanger S.
McDonnell Aircraft Corporation, *Project Mercury Bi-Monthly Engineering Status Report, August 1, 1959 to October 1, 1959*, p. 1.

September 3

Ground rules for prelaunch preparations were forwarded by the Space Task Group to McDonnell to serve as a guideline in the design of Mercury checkout equipment. Items covered included blockhouse equipment, checkout trailer, and telemetry trailer.
Letter, Space Task Group to Logan T. MacMillan, Project Director, McDonnell Aircraft Corporation, subject: Ground Support Equipment, Sept. 3, 1959.

September 9

A Big Joe Atlas boilerplate Mercury spacecraft model (fig. 24) was successfully launched and flown from Cape Canaveral, although booster-engine separation did not occur. Objectives of this test flight were to determine the performance of the ablation shield and measure afterbody heating; to determine the flight dynamics of the spacecraft during reentry; to evaluate the adequacy of the spacecraft recovery system and procedures; to familiarize operating personnel with Atlas launch procedures; to evaluate loads on the spacecraft while in the flight environment; to observe and evaluate the operation of the spacecraft control system; and to recover the spacecraft. The flight was considered to be highly successful, and a majority of the test objectives were attained. The heat shield temperatures (reaching a peak of 3,500 degrees F) were below those expected, but were close enough to provide data for the engineering design of the Mercury heat shield. Space Task Group officials were also pleased that the spacecraft could reenter the atmosphere at high angles-of-attack and maintain its heat shield in a forward position without using the control system. The spacecraft was picked up by the recovery force about 8 hours after lift-off. Because of the success of this flight, a similar launch was considered unnecessary and accordingly was canceled.
Memo, George Low to NASA Administrator, subject: Big Joe Shot, September 9, 1959; Preliminary Data, subject: Noise Environments for Big Joe I Test Vehicle, undated.

The Space Task Group provided McDonnell with guidance in the development of the "Astronauts' Handbook." Topics included such items as a descriptive resume of normal and emergency procedures to be followed on the check lists. The book was divided into three sections: "The Normal Operational Procedures," "The Emergency Operational Procedures," and "The Failure Analysis Procedures."

> Letter, Paul E. Purser, Space Task Group, to Logan T. MacMillan, Project Director, McDonnell Aircraft Corporation (no subject), Sept. 9, 1959.

Figure 24. Big Joe on launch pad at Cape Canaveral for ballistic reentry flight test.

September 10-11

At a spacecraft mock-up review, the astronauts submitted several recommended changes which involved a new instrument panel (fig. 25), a forward centerline window, and an explosive side egress hatch.

> McDonnell Aircraft Corporation, *Project Mercury Bi-Monthly Engineering Status Report, August 1, 1959 to October 1, 1959*, p. 23.

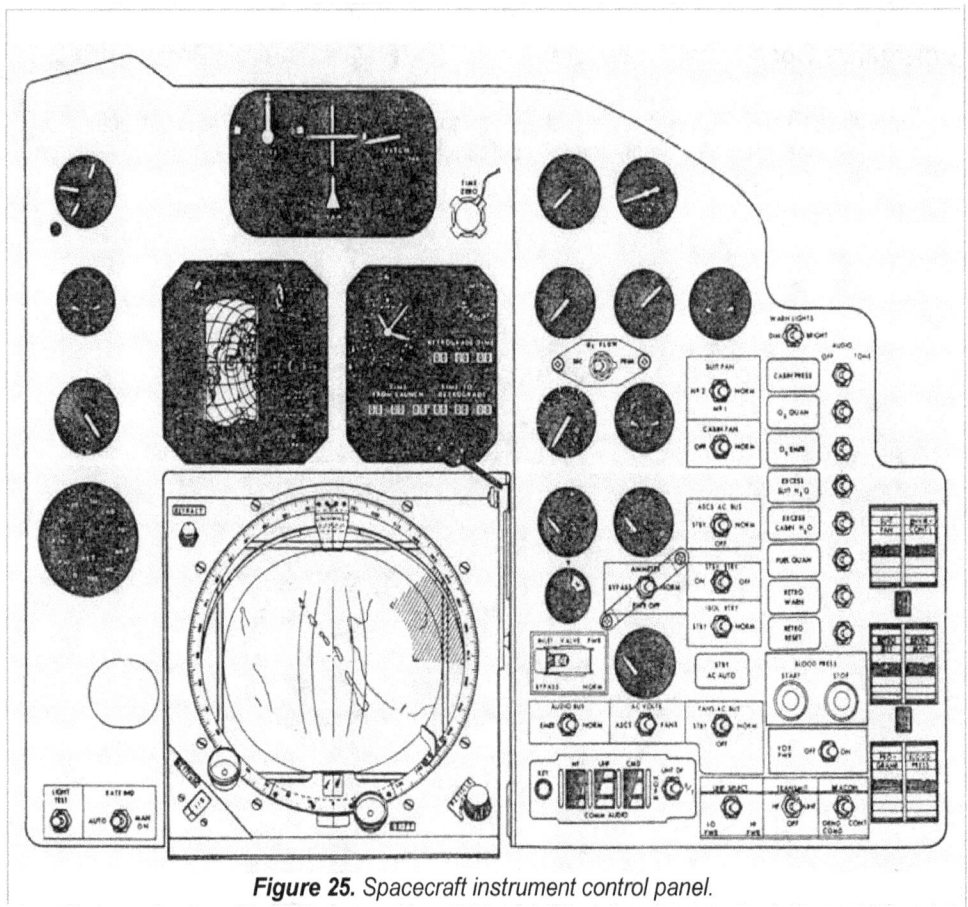

Figure 25. Spacecraft instrument control panel.

September 11

After a preliminary study of the Mercury environment with regard to astronaut food and water requirements, Dr. Douglas H. K. Lee estimated that water use would be in the order of 500 cu cm/hr and that the caloric intake per day would be about 3,200 calories of food. Dr. Lee was a member of the Natick Quartermasters Research and Engineering Laboratory.
> Memo, Dr. W. S. Augerson, Life Systems Branch, to Chief Flight Systems Division, Space Task Group, subject: Trip Report, Sept. 11, 1959.

September 15

Walter C. Williams was appointed Associate Director for Project Mercury Operations, and also the prime NASA-Department of Defense contact for Mercury flight operations.
> Information supplied from Personnel Records by Kathryn Walker, Personnel Division, Manned Spacecraft Center, March 1963.

September 16

The Langley Research Center was in the process of conducting ablation heat-shield tests on nine model shields in support of Project Mercury. However, the Big Joe test of the week before demonstrated the feasibility of the ablation heat-shield concept for reentry and verified the suitability of the materials selected for such purposes.
> Memo, Robert L. O'Neal to Chief, Flight Systems Division, Space Task Group, subject: Progress to Date on Ablation Tests in Support of Project Mercury, Sept. 16, 1959.

September 19

An air launch of a Mark II parachute (drogue) test vehicle was conducted by the NASA Flight Research Center. This test, the 15th in the series, concluded the Project Mercury drogue parachute development and qualification tests.
> Memo, Flight Research Center to NASA Hq., subject: Final Project Mercury Status Report, Sept. 19, 1959, Oct. 5, 1959, with enclosures.

September 21

Between this date and October 10, 1959, a research program was carried out by the Aviation Medical Acceleration Laboratory to measure the effects of sustained acceleration on the pilot's ability to control a vehicle. Various side-arm controllers were used, and it appeared that the three-axis type (yaw, roll, and pitch) was the most satisfactory. (See fig. 26.) Later this configuration was extensively evaluated and adopted for use in the control system of the Mercury spacecraft.
> Memo, Brent Y. Creer and Rodney C. Wingrove to Director, NASA Ames, subject: Preliminary Results of Pilot's Side-Arm Controller Tests Conducted on the AMAL-NAOC Centrifuge, Johnsville, Pennsylvania, February 26, 1960.

September 22

A paper was issued covering "Results of Studies Made to Determine Required Retrorocket Capability." The intent of this study was to provide for pilot safety for landing during any emergency condition, as well as at the end of a normal mission.
> NASA Project Mercury Working Paper 102, Sept. 22, 1959.

September (during the month)

An operational analysis study report of possible recovery forces required for a three-orbit Mercury mission was received from the Grumman Aircraft Engineering Corporation. By using this document, the Space Task Group was continuing to refine recovery requirements for all Mercury flights. This work involved the development of a satisfactory helicopter recovery technique and the conduct of tests to determine optimum spacecraft location aids.

NASA Space Task Group, *Project Mercury [Quarterly] Status Report No. 4 for Period Ending October 31, 1959.*

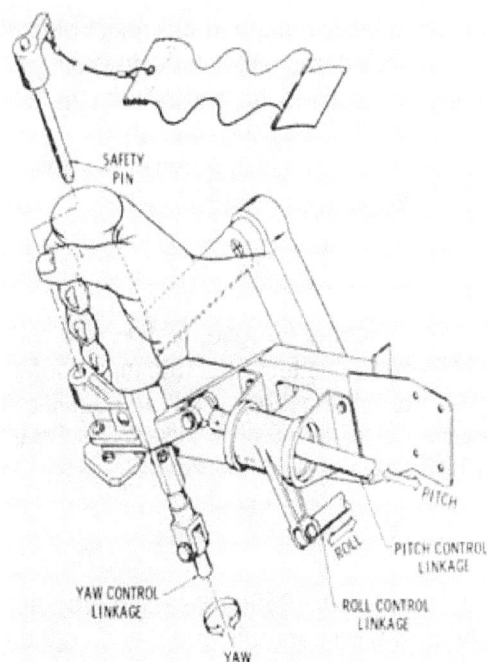

Figure 26. *Three-axis hand controller.*

October 1

On this date, funds were approved by NASA Headquarters for the following major changes to the Mercury spacecraft: egress hatch installation (CCP-58-1), astronaut observation window installation (CCP-73); rate stabilization and control system (CCP-61-2), main instrument and panel redesign (CCP-76), installation of reefed ringsail landing parachute (CCP-41), and nonspecification configurations of spacecraft (CCP-8). With reference to the last item, the original contract with McDonnell had specified only one spacecraft configuration, but the various research and development flight tests required changes in the configuration.
> Memo, George Low to NASA Director of Space Flight Development, subject: Budgetary Approval of Proposed Project Mercury Procurement, Oct. 1, 1959.

October 2

Specifications for the Mercury pressure suit were issued. The suit procurement program was divided into two phases: Phase I, operational research suits which could be used for astronaut training, system evaluation, and further suit development; and Phase II, Mercury pressure suits in the final configuration.
> Memo, Space Task Group to Langley Research Center, Attn: Procurement Officer, subject: Project Mercury Astronaut Pressure Suit Procurement, Oct. 2, 1959, with enclosures.

October 4

A Little Joe launch vehicle carrying a boilerplate spacecraft (LJ-6) was successfully launched from Wallops Island. Objectives of the flight were to check the integrity of the launch vehicle airframe and motor system, check the operations of the launcher, to check the validity of the calculated wind corrections, to obtain performance and drag data, and to check the operation of the destruct system. The flight, lasting 5 minutes 10 seconds, gained a peak altitude of 37.12 statute miles, and a range of 79.4 statute miles. The destruct packages carried on board the Little Joe launch vehicles were successfully initiated well after the flight had reached its apex. There was a slight malfunction in the Little Joe launch vehicle when ignition of the two second-stage Pollux motors fired before the exact time planned. Actually, the planned trajectory was little affected and the structural test of the vehicle, really greater than planned, was benefited.
> NASA Release No. 59-235, subject: NASA Conducts Little Joe Test Launch, Oct. 4, 1959; Memo, George Low to NASA Administrator, subject: First Little Joe Launching, Oct. 5, 1959.

October 13

Explorer VII achieved orbit on this date and began providing significant geophysical information on solar and earth radiation, magnetic storms, and micrometeorite penetration. This satellite also successfully demonstrated a method of controlling internal temperatures.
> Goddard Space Flight Center Chart, Satellites and Space Probe Projects as of July 1962.

October 15

Space Task Group personnel held a meeting at Langley with representatives from the Lewis Research Center to clarify Project Mercury research support needs at Lewis. During the course of discussion, several test and support areas were agreed upon. As an example, Lewis would conduct separation tests in which full-scale hardware was used to determine if a satisfactory separation existed. In these tests separation would occur when the posigrade rockets were fired after burnout of the Atlas during an ordinary mission. Lewis would seek to determine if there were any harmful effects due to flame impingement either on the Atlas booster or on the wiring of the retrograde package. In addition, Lewis would determine the actual effective impulse of the posigrade rockets during separation. Lewis also agreed to support Space Task Group in developing pilot techniques in a special tunnel at Lewis. The objectives were to determine a pilot's capability to stabilize spacecraft attitudes in space. Lewis had a large gimballed system in the tunnel that would simulate the motions of space conditions, but in a sea-level environment. It was thought, however, that experience in the gamballed system would be beneficial to the pilots. A third area of support involved retrorocket calibration tests. At that time, Space Task Group was concerned that when the retrorockets were fired, the spacecraft would be considerably upset while in orbital flight. Lewis would use its high-altitude tunnel at maximum capability to determine the extent of the upset and assist in devising means to control the situation. Lewis also agreed to check the hydrogen-peroxide-fueled control system to obtain starting and performance characteristics of the reaction jets. In the last area of this series of studies and tests, Lewis was to study the escape rocket plume when the rocket was fired at high altitudes to determine the effect of the plume on the spacecraft. It was believed that the plume would completely envelop the spacecraft.

> Memo, Maxime A. Faget, Chief, Flight Systems Division to Project Director, subject: Status of Test Work Being Conducted at the Lewis Research Center in Conjunction with Project Mercury, Oct. 22, 1959.

October 20

Requests were initiated to test the Mercury spacecraft afterbody shingles at the Navy's Dangerfield test facility for heat resistance and dynamic-pressure capabilities.

> Memo, George Low, Chief, Manned Space Flight to NASA Director of Space Flight Development, subject: Tests of Project Mercury Shingle Structure, Oct. 20, 1959.

October 30

A meeting of Space Task Group, Wallops Station, and McDonnell personnel was held to review and evaluate Mercury escape-system qualification-test results. In the continuing efforts of this activity, the responsibility in attaining test objectives was apportioned among the three organizations.

> Project Mercury, Minutes of Meeting, subject: Escape System Qualification Test, Oct. 30, 1959.

October (during the month)

North American Aviation and Minneapolis-Honeywell were notified to proceed with the production of hardware for an air-supplied launch-vehicle control system.

> NASA Space Task Group, *Project Mercury [Quarterly] Status Report No. 4 for Period Ending October 31, 1959.*

McDonnell received the first ablative heat shield (fig. 27), designated for installation on Spacecraft No. 1. This particular heat shield was based on the Big Joe design, and was manufactured by General Electric.

> NASA Space Task Group, *Project Mercury [Quarterly] Status Report No. 4 for Period Ending October 31, 1959.*

November 1

The "Handbook of Operation and Service Instructions, Horizon Scanner Test, Serial MDE 4590011" was published. This document was revised and reissued on June 6, 1960.

> McDonnell Report SEDR-120, *Handbook of Operation and Service Instructions, Horizon Scanner Test Set MDE 4590011, Contract NAS 5-59*, June 6, 1960.

Figure 27. Mercury spacecraft heat shield after recovery.

November 4

Little Joe 1-A (LJ-1A) was launched in a test for a planned abort under high aerodynamic load conditions. This flight was a repeat of the Little Joe (LJ-1) that had been planned for August 21, 1959 (escape rocket fired 31 min before the intended launch of the Little Joe launch vehicle). After lift-off, the pressure sensing system was to supply a signal when the intended abort dynamic pressure was reached (about 30 sec after launch). An electrical impulse was then sent to the explosive bolts to separate the spacecraft from the launch vehicle. Up to this point, the operation went as planned, but the impulse was also designed to start the igniter in the escape motor. The igniter activated, but pressure failed to build up in the motor until a number of seconds had elapsed. Thus the abort maneuver, the prime mission of the flight, was accomplished at a dynamic pressure that was too low. For this reason a repeat of the test was planned. All other events from the launch through recovery occurred without incident. The flight attained an altitude of 9 statute miles, a range of 11.5 statute miles, and a speed of 2,021.6 miles per hour.

> NASA Space Task Group, *Project Mercury [Quarterly] Status Report No. 5 for Period Ending January 31, 1960*.

November 5

The astronauts were fitted with pressure suits and indoctrinated as to use at the B. F. Goodrich Company, Akron, Ohio.
> Memo, L. N. McMillian to Chief, Flight Systems Division, Space Task Group, subject: Trip Report, Nov. 17, 1959.

November 8

Between this date and December 5, 1959, the tentative design and layout of the Mercury Control Center to be used to monitor the orbiting flight of the Mercury spacecraft were completed. The control center would have trend charts to indicate the astronaut's condition and world map displays to keep continuous track of the Mercury spacecraft. (See fig. 28.)
> Memo Howard S. Carter to Associate Director, Langley Research Center, subject: Progress of Nov. 8-Dec. 5, 1959, on Langley Support of Project Mercury, Dec. 8, 1959.

Figure 28. Mercury control center at Cape Canaveral.

November 10

Space Task Group personnel visited McDonnell to monitor the molding of the first production-type couch for the Mercury spacecraft.
> Memo, Gerald J. Pesman to Chief, Flight Systems Division, Space Task Group, subject: Visit to MAC to Monitor Molding of First Production-Type Couch, Nov. 10, 1959.

November 12

A NASA-Department of Defense agreement was signed by NASA Administrator T. Keith Glennan and Deputy Secretary of Defense Thomas Gates, relevant to the principles governing reimbursement of costs incurred by NASA or the Department of Defense in support of Project Mercury.
> NASA General Management Instruction 2-3-5, Attachment A, subject: Agreement Between the Department of Defense and the National Aeronautics and Space Administration Concerning Principles Governing Reimbursements of Costs, Nov. 12, 1959.

November 16-20

Wearing the Mercury pressure suits, the astronauts were familiarized with the expected reentry heat pulse at the Navy Aircrew Equipment Laboratory, Philadelphia, Pennsylvania.
> Memo, L. N. McMillian to Chief, Flight Systems Division, Space Task Group, subject: Trip Report, Nov. 20, 1959.

November 20

At the fifth Mercury Coordination Meeting, the Army Ballistic Missile Agency proposed the installation of an open-circuit television system in the Mercury-Redstone second and third flights (MR-2 and MR-3). The purpose of the system was to observe and relay launch vehicle and spacecraft separation data.
> Letter, Army Ballistic Missile Agency to NASA Administrator (no subject), March 2, 1960, with TV Proposal Enclosure.

November 27

The Arnold Engineering Development Center tested the Grand Central solid-fuel rocket motor used to propel the Mercury spacecraft escape system. The purpose of the test was to verify altitude ignition and to determine the combustion-chamber-pressure-time curve.
> *Chronology of the Arnold Engineering Development Center; History of Arnold Engineering Development Center, July-December 1959*, Vol. I, pp. 47-52.

The Air Force School of Aviation Medicine agreed to provide a biopack experiment for the Little Joe 2 flight. Included in the pack were track plates of barley, nerve cells from a rat, tissue culture, and other specimens of that type.

> Memo, G. D. Smith, NASA Manned Space Flight, to files, subject: Biopack Little Joe No. 2, Nov. 30, 1959.

November (during the month)

The first manned development system tests were completed at the AiResearch Manufacturing Division, Garrett Corporation. Tests were conducted in the altitude chamber to determine proper functioning of all system valves and components. A McDonnell subject was clothed in a Mercury-type presure suit for these tests. Preliminary data from these tests indicated that the system functioned satisfactorily.

    NASA Space Task Group, *Project Mercury [Quarterly] Status Report No. 5 for Period Ending January 31, 1960.*

Between November 1959 and January 1960, 10 developmental full-pressure suits were delivered to the astronauts and other subjects. These suits were used in various Mercury training and development programs. (See Oct. 2, 1959 entry). Several problem areas were denoted. One involved stretching which complicated the suit mobility problem. This matter was being investigated, and one of the solutions was felt to be undersizing to allow for a suit growth factor. In addition, modifications would have to be made in suit insulation to provide for better pilot mobility. These problems were to be expected in a developmental program.

    NASA Space Task Group, *Project Mercury [Quarterly] Status Report No. 5 for Period Ending January 31, 1960.*

Between November 1959 and January 1960, the general design of the Mercury couch was completed, and couches were molded for the astronauts and medical personnel associated with the program. (See fig. 29.)

    NASA Space Task Group, *Project Mercury [Quarterly] Status Report No. 5 for Period Ending January 31, 1960.*

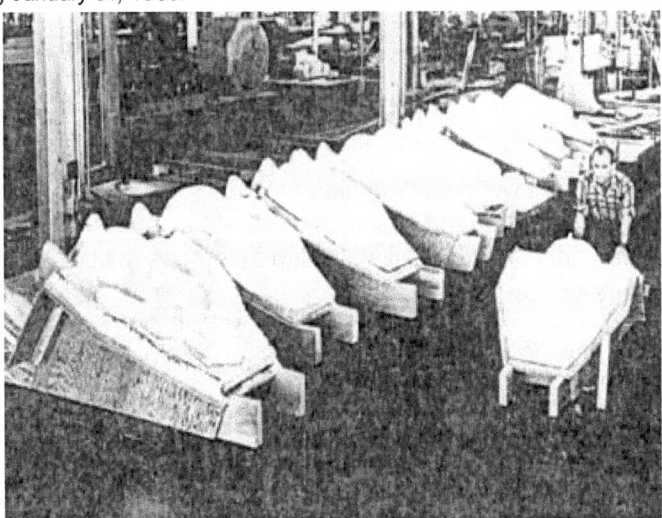

Figure 29. Plaster forms used in construction of research and development versions of contour couches.

December 4

Little Joe 2 (LJ-2) was launched from Wallops Island to determine the motions of the spacecraft escape tower combination during a high-altitude abort, entry dynamics without a control system, physiological effects of acceleration on a small primate, operation of the drogue parachute, and effectiveness of the recovery operation. Telemetry was set up to record some 80 bits of information on the flight. The abort sequence was initiated by timers after 59 seconds of elapsed flight time at an altitude of about 96,000 feet and a speed of Mach 5.5. Escape motor firing occurred as planned and the spacecraft was whisked away at a speed of about Mach 6 to an apogee of 53.03 statute miles. All other sequences operated as planned, and spacecraft recovery was effected in about 2 hours from lift-off. The primate passenger, "Sam," an American-born rhesus monkey, withstood the trip and the recovery in good condition. All objectives of the mission were met.

> Memo, George Low to NASA Administrator, subject: Little Joe Test No. 3, (LJ-2), December 5, 1959; NASA Space Task Group, *Project Mercury [Quarterly] Status Report No. 5 for Period Ending January 31, 1960.*

December 7

Tenney Engineering Corporation was chosen by the Space Task Group to construct the Mercury altitude test chamber in Hanger S at Cape Canaveral. When completed, altitude pressure would simulate 225,000 feet. The chamber, a vertical cylinder with domed ends, was 12 feet in diameter and 14 feet high. The chamber was designed to allow a partial spacecraft functional check in a near-vacuum environment.

> Memo, Warren J. North, Chief, Manned Satellite, to Director, Space Flight Development, subject: Request for Approval of Project Mercury Altitude Test Facility, Dec. 8, 1959.

December 8

Two Thiokol retrorockets for the Mercury spacecraft were tested at the Arnold Engineering Development Center engine test facility. The test objectives were to evaluate ignition characteristics.

> NASA Space Task Group, *Project Mercury [Quarterly] Status Report No. 5 for Period Ending January 31, 1960; History of Arnold Engineering Development Center, July-December 1959,* Vol. I, pp. 47-52.

December 22

The Redstone launch vehicle for the first Mercury-Redstone mission (MR-1) was installed on the interim test stand at the Army Ballistic Missile Agency for static testing.

> NASA Space Task Group, *Project Mercury [Quarterly] Status Report No. 5 for Period Ending January 31, 1960.*

December 31

Thrust cut-off sensor reliability and qualification tests were accepted, because of the similarity to Lockheed functional environmental evaluation tests of similar units used in the Polaris program. This component, fabricated by the Donner Scientific Company, was accepted by NASA.

> NASA Space Task Group, *Project Mercury [Quarterly] Status Report No. 5 for Period Ending January 31, 1960.*

At the end of the year, NASA funds in support of Project Mercury had been obligated to the listed organizations as follows: Air Force Ballistic Missile Division, NASA Order HS-36, Atlas launch vehicles, $22,830,000; Army Ordnance Missile Command, NASA Order HS-44, Redstone launch vehicles, $16,060,000; and McDonnell Aircraft Corporation, NASA Order 5-59, Mercury spacecraft, $49,407,540.

> Memo, Glenn F. Bailey to Director of Project Mercury, subject: Obligation of Funds, December 31, 1959.

Since being awarded the Mercury contract, McDonnell had expended 942,818 man-hours in engineering; 190,731 man-hours in tooling; and 373,232 man-hours in production.

> Letter, McDonnell Aircraft Corporation to Space Task Group, subject: Contract NAS 5-59, Monthly Financial Report, Jan. 22, 1960.

The Mercury astronauts completed basic and theoretical studies of Project Mercury in their training program and began practical engineering studies. This phase of the program was designed to provide a background in basic astronautical sciences, and included such subjects as "Space Climate" and "Astronomy of the Universe." Shortly thereafter the astronauts began a practical training program involving egress training, methods of arresting rapid spacecraft motions, and familiarization with the weightless conditions of space flight.

> NASA Space Task Group, *Project Mercury [Quarterly] Status Report No. 5 for Period Ending January 31, 1960.*

December (during the month)

A weightless flying training program was started by the Mercury astronauts in the F-100 aircraft at Edwards Air Force Base, California. Eating, drinking, and psychomotor tests were conducted while the astronauts were in a weightless state.
> NASA Space Task Group, *Project Mercury [Quarterly] Status Report No. 5 for Period Ending January 31, 1960.*

The Space Task Group approved monitoring facilities proposed by the Stromberg-Carlson Division for the Mercury Control Center at Cape Canaveral and Bermuda.

> NASA Space Task Group, *Project Mercury [Quarterly] Status Report No. 5 for Period Ending January 31, 1960.*

In the development of the Mercury spacecraft reaction control system, Bell Aircraft Corporation started the preliminary flight rating test of the automatic subsystem. (See fig. 30.)

NASA Space Task Group, *Project Mercury [Quarterly] Status Report No. 5 for Period Ending January 31, 1960.*

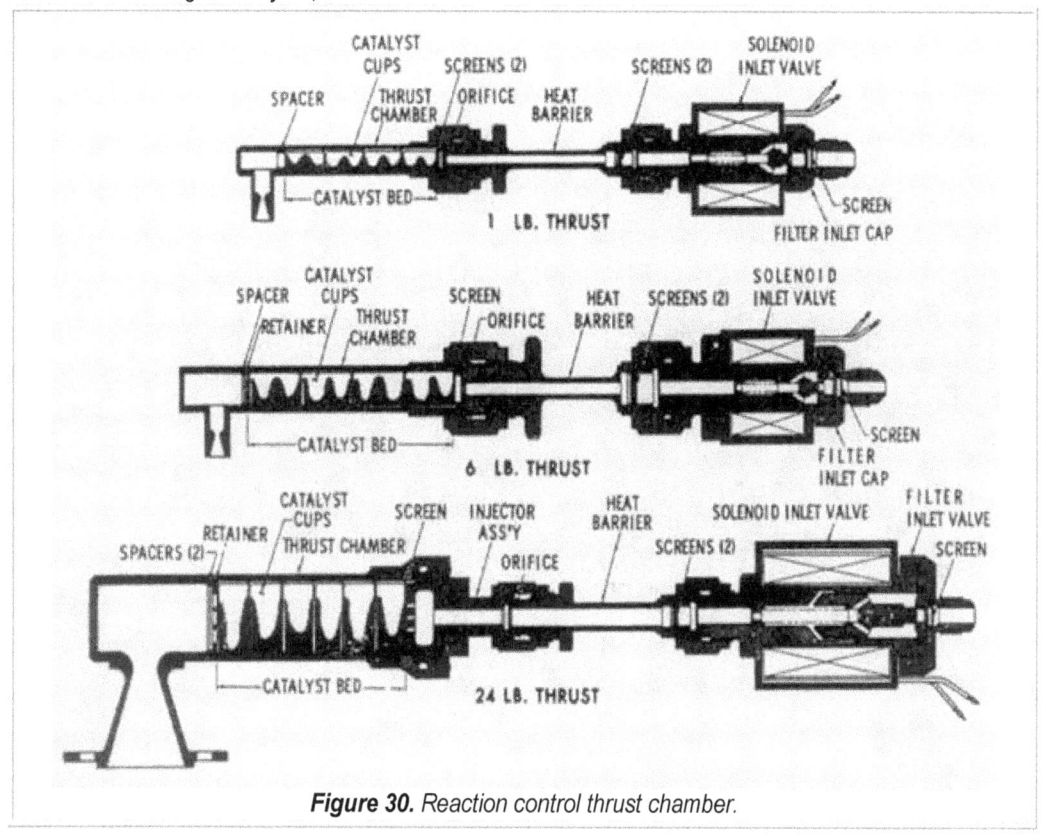

Figure 30. Reaction control thrust chamber.

PART II (B)
Research and Development Phase of Project Mercury
January 1960 through May 5, 1961

1960

January 6

The Project Mercury data reduction plan was approved. Space Task Group's study entitled "Semi-Automatic Data Reduction" had been completed and submitted to NASA Headquarters for review on December 21, 1959.
> Space Task Group Study, Semi-Automatic Data Reduction, Dec. 21, 1959, with endorsement.

January 11

A contract (NAS 1-430) was signed by NASA and the Western Electric Company in the amount of $33,058,690 for construction and engineering of the Mercury tracking network.
> Chart, Contract for Mercury Tracking Network, Summary Cost by Item and Team Members, Chart undated; NASA Langley Report, subject: Status Report, Project Mercury Tracking and Ground Instrumentation System transmitted to NASA Headquarters, March 17, 1960.

January 15

A document entitled "Overall Plan for Department of Defense Support for Project Mercury Operations" was reviewed and approved by NASA Headquarters and the Space Task Group.
> NASA Space Task Group, *Project Mercury [Quarterly] Status Report No. 5 for Period Ending January 31, 1960.*

Based on requirements listed in Space Task Group Working Paper No. 129, covering the Project Mercury recovery force, the Navy issued "Operation Plan COMDESFLOTFOUR No. 1-60." This plan provided for recovery procedures according to specified areas and for space recovery methods. Procedures for Mercury-Redstone and Mercury-Atlas missions were covered.

> Navy Ops Plan 1-60, subject: NASA Statement of Recovery Requirements for Orbital Flights, Jan. 15, 1960.

Qualification tests on a programmer fabricated by the Wheaton Engineering Company for Project Mercury were started and completed by March 28, 1960.

NASA Space Task Group, *Project Mercury [Quarterly] Status Report No. 6 for Period Ending April 30, 1960.*

January 18

Walter C. Williams proposed the establishment of a Mercury-Redstone Coordination Committee to monitor and coordinate activities related to Mercury-Redstone flight tests.

> Letter, Walter C. Williams, Associate Director for Project Mercury, to Dr. Kurt H. Debus, Director, Missile Firing Laboratory, subject: Proposal for Mercury-Redstone Coordination Committee, Jan. 18, 1960.

A proposal was made by Walter C. Williams, Associate Director of Project Mercury Operations, that the Mercury-Atlas flight test working group become an official and standing coordination body. This group brought together representation from the Space Task Group, Air Force Ballistic Missile Division, Convair Astronautics, McDonnell Aircraft Corporation, and the Atlantic Missile Range. Personnel from these organizations had met informally in the past on several occasions.

> Letter, Walter C. Williams, Associate Director of Project Mercury, to Major General Donald N. Yates, Department of Defense Representative, Project Mercury Support Operations, subject: Mercury-Atlas Flight Test Working Group, Jan. 18, 1960.

January 19

In keeping with a concept of using certain off-the-shelf hardware items that were available for the manufacture of Project Mercury components, companies around London, England, were visited throughout 1959. Potential English vendors of such items as the SARAH beacon batteries (later chosen), miniature indicators, time delay mechanisms, hydrogen-peroxide systems, and transducers were evaluated. A report of the findings was submitted on the cited date.

> Memo, Thomas V. Chambers to Chief of Flight Systems Division, Space Task Group, subject: Visit to Companies in United Kingdom, January 19, 1960.

January 21

At a meeting to draft fiscal year 1962 funding estimates, the total purchase of Atlas launch vehicles was listed as 15, and the total purchase of Mercury spacecraft was listed as 26.

> NASA Headquarters Memo to File by John Disher, Advanced Manned Systems, subject: Preliminary Estimates of FY 1962 Funding Requirements for Project Mercury, Jan. 21, 1960.

Little Joe 1-B (LJ-1B) was launched from Wallops Island with a rhesus monkey, "Miss Sam," aboard. (See fig. 31.) Test objectives for this flight were the same as those for Little Joe 1 (LJ-1) in which the escape tower launched 31 minutes before the planned launch, and Little Joe 1-A (LJ-1A), wherein the dynamic buildup in the abort maneuver was too low. A physiological study of the primate, particularly in areas applying to the effects of the rapid onset of reverse acceleration during abort at maximum dynamic pressure, was also made. In addition, the Mercury helicopter recovery system was exercised. During the mission, all sequences operated as planned; the spacecraft attained a peak altitude of 9.3 statute miles, a range of 11.7 statute miles, and a maximum speed of 2,021.6 miles per hour. Thirty minutes from launch time, a Marine recovery helicopter deposited the spacecraft and its occupant at Wallops Station. "Miss Sam" was in good condition, and all test objectives were successfully fulfilled.

> Memo, George Low to NASA Administrator, subject: Little Joe 1-B (Test No. 4), January 21, 1960, January 22, 1960; NASA Space Task Group, *Project Mercury [Quarterly] Status Report No. 5 for Period Ending January 31, 1960.*

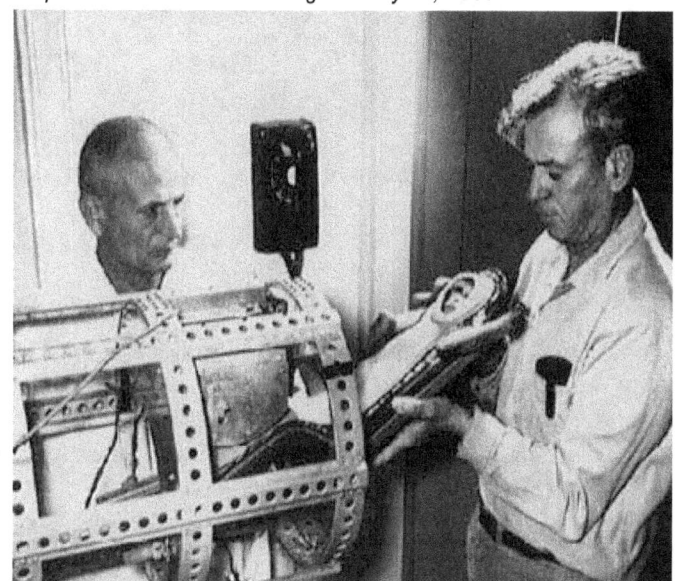

Figure 31. Rhesus monkey, "Miss Sam," being placed in container for LJ-1B flight.

January 25

McDonnell delivered the first production-type Mercury spacecraft to the Space Task Group at Langley in less than 1 year from the signing of the formal contract. (See fig. 32.) This spacecraft was a structural shell and did not contain most of the internal systems that would be required for manned space flight. After receipt, the Space Task Group instrumented the spacecraft and designated it for the Mercury-Atlas 1 (MA-1) flight.

> NASA Space Task Group, *Project Mercury [Quarterly] Status Report No. 5 for Period Ending January 31, 1960.*

Figure 32. Manufacture of Mercury spacecraft at McDonnell plant, St. Louis, Mo.

January 31

Six chimpanzees were rated as being trained and ready to support Mercury-Redstone or Mercury-Atlas missions. Other chimpanzees were being shipped from Africa to enter the animal training program.
> NASA Space Task Group, *Project Mercury [Quarterly] Status Report No. 5 for Period Ending January 31, 1960.*

January (during the month)

Specifications for equipment and systems to be used for the training of the remote-site flight controllers and Mercury control center operations personnel were forwarded to the Western Electric team. The remote-site training was divided into two stages: off-range and on-range. The off-range portion consisted of practice runs on a typical set of controllers' consoles tied into an astronaut procedures trainer. The on-range part was planned at two stations within the United States and from here, controllers would be assigned to tracking sites for full range rehearsals and a mission.
> NASA Space Task Group, *Project Mercury [Quarterly] Status Report No. 5 for Period Ending January 31, 1960.*

NASA presented its basic communications requirements for Project Mercury to Western Electric, and the Company's interim proposal to satisfy these requirements was accepted in February 1960.

> NASA Space Task Group, *Project Mercury [Quarterly] Status Report No. 5 for Period Ending January 31, 1960.*

Qualification tests were completed on the Mercury spacecraft pilot cameras and instrument viewing cameras.

> NASA Space Task Group, *Project Mercury [Quarterly] Status Report No. 5 for Period Ending January 31, 1960.*

February 1

Qualification tests of the Mercury spacecraft periscope were completed.
> NASA Space Task Group, *Project Mercury [Quarterly] Status Report No. 5 for Period Ending January 31, 1960.*

A study was completed on the "External and Internal Noise of Space Capsules." This study covered the acoustic environments of missile and space vehicles including noise generated by the rocket engines, air-boundary layers, and on-board equipment. Data used included noise measurements compiled from the Big Joe I and Little Joe 2 flight tests. These tests were a part of the internal and external noise study that had been in progress since early 1959. NASA officials were still of the opinion that the internal noise level was too high for pilot comfort. Space Task Group felt that data were needed on noise transmission through an actual production-model spacecraft structure.

> William H. Mayer and David A. Hilton, subject: External and Internal Noise of Space Capsules, Feb. 1, 1960; Memo, Harvey H. Hubbard to Associate Director, NASA Langley, subject: Noise Measurements of Big Joe and Little Joe Mercury Vehicles, Feb. 17, 1960.

February 5

A meeting was held to relay the decision that beryllium shingles would be used as the best heat protection material on the cylindrical section of the Mercury spacecraft.
> Minutes of Meeting, Space Task Group, subject: Meeting at MAC on Beryllium Shingles, Feb. 11, 1960.

Final design approval test of the Mercury telemetry equipment was completed, and reliability test of this equipment was completed on February 27, 1960.

> NASA Space Task Group, *Project Mercury [Quarterly] Status Report No. 7 for Period Ending July 31, 1960.*

Colonel George M. Knauf of the Air Force Surgeon General's office began the compilation of a medical-monitor training program in support of Project Mercury. The aims of this program were to brief the monitors on medical problems in space prior to their participation in support of Mercury flights. Colonel Knauf is now a member of NASA Headquarters Manned Space Flight Office.

> Memo, Dr. Stanley C. White, Head, Life Systems Branch to Chief, Flight Systems Division, Space Task Group, subject: Trip to USAF Surgeon General's Office,

Washington, D.C., on February 5, 1960, to Discuss Early Training of Medical Monitors with Colonel Knauf, Feb. 8, 1960.

February 8

Tests were started by the Army Ballistic Missile Agency for the mission abort sensing program to be integrated in the Mercury-Redstone phase of Project Mercury.

>Memo, Jack C. Heberlig to Chief of Flight Systems Division, Space Task Group, subject: Mercury-Redstone Coordination Visit to ABMA on Feb. 10, 1961, Feb. 15, 1960.

February 11

Responsibilities of the Mercury launch coordination office were specified by the Space Task Group. A few of the listed duties included responsibilities associated with Department of Defense support; overall coordination of launch activities; compilation of information related to launch support requirements; and representing Mercury at Atlas or Redstone Flight Test Group meetings. Walter C. Williams made a proposal for an activity along these lines on January 18, 1960.

>Memo, Walter C. Williams to Space Task Group Staff, subject: Responsibilities for Mercury Launch Coordination Office, Feb. 11, 1960.

February 12

With Project Mercury about to enter a heavy operational phase, an operations coordination group was established at the Atlantic Missile Range. Christopher C. Kraft, Jr. was appointed to head this group.

>Memo, Walter C. Williams to Space Task Group Staff, subject: Organization for Mercury Field Operations, Feb. 12, 1960.

February 15

Mercury spacecraft battery qualification tests were completed.

>NASA Space Task Group, *Project Mercury [Quarterly] Status Report No. 6 for Period Ending April 30, 1960.*

Mercury landing system and post-landing equipment tests were completed. (See fig. 33.)

>NASA Space Task Group, *Project Mercury [Quarterly] Status Report No. 6 for Period Ending April 30, 1960.*

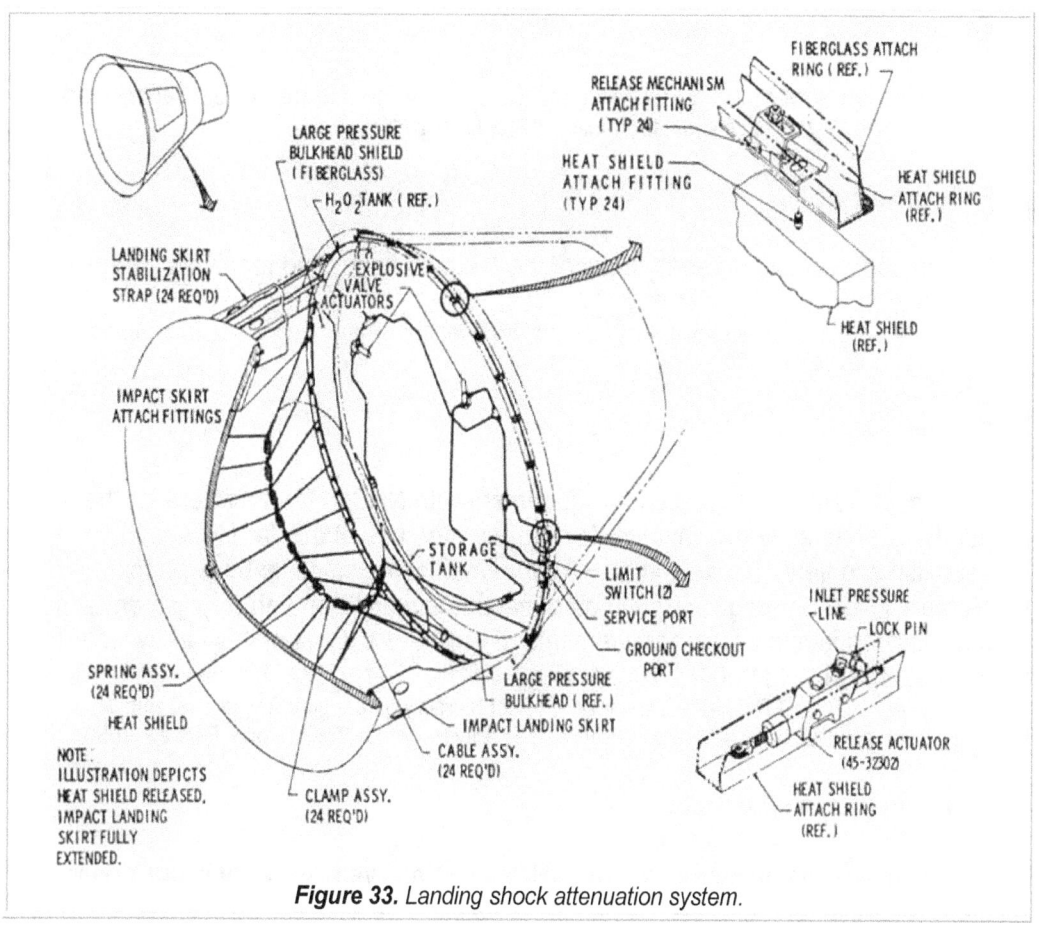

Figure 33. Landing shock attenuation system.

February 18

Mercury remote-site flight controllers were appointed, and training was inaugurated by a series of Space Task Group lectures that covered facilities, network systems, operations, and other details. In addition, a program was established for familiarization, orientation, and specialized instruction of the Department of Defense group of aeromedical staff personnel designated as members of flight controller teams.
> NASA Space Task Group, *Project Mercury [Quarterly] Status Report No. 6 for Period Ending April 30, 1960.*

February 22

Tests were completed on the Mercury spacecraft automatic stabilization and control system.
> NASA Space Task Group, *Project Mercury [Quarterly] Status Report No. 5 for Period Ending January 31, 1960.*

February 26

The establishment of a Project Mercury tracking site in Australia was sanctioned.
    Emme, *Aeronautics and Astronautics: 1915-1960*, p. 120.

February 27

Design approval and reliability tests of the Mercury command receivers were completed.
    NASA Space Task Group, *Project Mercury [Quarterly] Status Report No. 7 for Period Ending July 31, 1960.*

February 29

The Space Task Group placed a requirement with NASA Headquarters for the purchase of an analog computing facility. Planned use of this facility was to establish and verify Mercury system requirements; it also could be used for Mercury follow-on programs such as a manned circumlunar vehicle program and other outer space program requirements of this nature. Cost of this facility was estimated to be $424,000.
    Memo, Robert R. Gilruth, Director of Project Mercury, to Langley Research Center, subject: Purchase of Analog Computing Facility for Space Task Group, Feb. 29, 1960.

February (during the month)

As part of their training program, the astronauts received 2 days of instruction in star recognition and celestial navigation presented by Dr. James Balten at the Morehead Planetarium in Chapel Hill, North Carolina. The purpose of this training was to assist the astronaut in correcting spacecraft yaw drifts. Practical experience was gained in this task by using a motorized trainer that simulated the view of the celestial sphere through the spacecraft observation window.
    NASA Space Task Group, *Project Mercury [Quarterly] Status Report No. 6 for Period Ending April 30, 1960.*

February-April

Agreements were signed with two Spanish firms to provide communications at the Grand Canary Island Mercury tracking site.
    NASA Space Task Group, *Project Mercury [Quarterly] Status Report No. 6 for Period Ending April 30, 1960.*

The Navy's School of Aviation Medicine modified a standard 20-man raft in such a way that it could be placed around the base of a floating spacecraft with impact skirt extended. When the device was inflated, the spacecraft rode high enough in the water to permit easy egress from the side hatch.

    NASA Space Task Group, *Project Mercury [Quarterly] Status Report No. 6 for Period Ending April 30, 1960.*

March 7-10

An indoctrination program in free-floating during weightless flight was conducted for the astronauts at the Wright Air Development Center. (See fig. 34.) The rear end of a C-131B aircraft was cleared and padded. Some 90 parabolas of 12 to 15 seconds of weightlessness each were flown. The objective was to present orientation problems of floating in space with the eyes opened and closed. Also, the astronauts made attempts to use tools and move weights while they were in a weightless condition.

NASA Space Task Group, *Project Mercury [Quarterly] Status Report No. 6 for Period Ending April 30, 1960.*

Figure 34. Astronauts in weightless flight in C-131 aircraft.

March 9

Position titles for Project Mercury operational flights were issued. During the flights, 15 major positions were assigned to Mercury Control Center, 15 in the blockhouse and 2 at the launch pad area. The document also specified the duties and responsibilities of each position.

Letter, Walter C. Williams, Associate Director, Project Mercury, to Major General Donald N. Yates, Department of Defense Representative, Project Mercury Support Operations, subject: Position Titles for Operations of Project Mercury, March 9, 1960.

March 11

Pioneer V, launched as a probe of the space between Earth and Venus, began to provide invaluable information on solar flare effects, particle energies and distributions and magnetic phenomena. Pioneer V continued to transmit such data until on June 26, 1960, when at a distance of 22.5 million miles from Earth, it established a new communications record.

Goddard Space Flight Center Chart: Satellites and Space Probe Projects as of July 1962.

The initial payment was made to the Australian Government by the Chase National Bank, New York City, on behalf of the National Aeronautics and Space Administration for support of the Mercury network.

>NASA Langley Report, subject: Status Project Mercury Tracking and Ground Instrumentation System, transmitted to NASA Headquarters, March 17, 1960.

March 16

The Space Task Group published recovery requirements for the Mercury-Atlas 1 (MA-1) flight test.

>Letter, Walter C. Williams, Space Task Group, to Commander, Destroyer Flotilla Four (no subject), Mar. 16, 1960.

March 19

An agreement between the United States and Spain on the Project Mercury tracking station in the Canary Islands was announced.

>Emme, *Aeronautics and Astronautics: 1915-1960*, p. 121.

March 28

Between this date and April 1, 1960, the astronauts received their first open-water egress training in the Gulf of Mexico off the coast of Pensacola, Florida, in cooperation with the Navy's School of Aviation Medicine. The training was conducted in conditions of up to 10-foot swells, and no problems were experienced. The average egress time was about 4 minutes from a completely restrained condition in the spacecraft to being in the life raft.

>Memo, Dr. W. S. Augerson to Chief, Flight Systems Division, Space Task Group, subject: Trip to B. F. Goodrich, Akron and NAS, Pensacola, March 29 thru April 2, 1960, April 6, 1960.

March 29

A decision was made by NASA Headquarters that the spacecraft prelaunch operation facility at Huntsville, Alabama, was no longer required. Spacecraft that were designated for Mercury-Redstone missions were to be shipped directly from McDonnell to Cape Canaveral, thereby gaining approximately 2 months in the launch schedule.

>Memo, Abe Silverstein, Director of Space Flight Program, NASA Headquarters to Director, Marshall Space Flight Center, subject: Mercury Capsule Prelaunch Operations at Huntsville, March 29, 1960.

March (during the month)

Qualification tests were started on the escape tower rocket. These tests were completed at the end of July 1960. As a part of the qualification program, three escape-rocket motors were successfully fired on a spacecraft model at conditions

corresponding to approximately 100,000 feet altitude in the Lewis Research Center altitude wind tunnel. One motor was tested on a four-component balance system to determine thrust misalignment of the rocket motor. According to test results, the rocket motor appeared to meet operational requirements.
> NASA Space Task Group, *Project Mercury [Quarterly] Status Report No. 6 for Period Ending April 30, 1960.*

The Secretary of Defense and the Joint Chiefs of Staff approved the "Overall Plan for Department of Defense Support for Project Mercury Operations" submitted by their representative, Major General Donald N. Yates. Following this decision, the Space Task Group prepared a series of documents to establish the required operations support. One was an "Operations Prospectus" which set forth the management techniques by which NASA planned to discharge its overall program responsibility in the operations area. A second was a "Programs Requirements Document" directed toward continuing operational support.
> NASA Space Task Group, *Project Mercury [Quarterly] Status Report No. 6 for Period Ending April 30, 1960.*

March-April

The Mercury-Atlas working panels were reorganized into four groups: coordination, flight test, trajectory analysis, and change control. Each panel was composed of at least one representative from NASA (Space Task Group), McDonnell, Air Force Ballistic Missile Division, Space Technology Laboratory, and Convair-Astronautics.
> NASA Space Task Group, *Project Mercury [Quarterly] Status Report No. 6 for Period Ending April 30, 1960.*

April 1

The first McDonnell production spacecraft was delivered to NASA at Wallops Island for the beach-abort test.
> Data Supplied by Ken Vogel, Mercury Project Office, Manned Spacecraft Center, April 9, 1963.

April 5

The Space Task Group notified the Ames Research Center that preliminary planning for the modification of the Mercury spacecraft to accomplish controlled reentry had begun, and Ames was invited to participate in the study. Preliminary specifications for the modified spacecraft were to be ready by the end of the month. This program was later termed Mercury Mark II and eventually Project Gemini.
> Memo, Charles J. Donlan, Associate Director of Project Mercury, to Ames Research Center, subject: Invitation to Participate in Preparing Specifications and Evaluation of Proposals for a Reentry Guidance System for Lifting Mercury, April 5, 1960.

April 7

Ablation tests on nine Mercury heat shield models in the subsonic arc tunnel at the Langley Research Center were completed. (See Sept. 16, 1959.)
> Letter, NASA Space Task Group to Logan T. McMillian, Project Manager, McDonnell Aircraft Corporation, subject: Ablation Tests Carried Out at Langley Research Center, April 7, 1960.

April 8

Construction of an altitude facility chamber to simulate space environment was completed in Hanger S at Cape Canaveral. The purpose of this facility was for spacecraft checkout and astronaut training. (See fig. 35.) Acceptance tests for this installation were completed on July 11, 1960. (See Dec. 7, 1959.)
> NASA Space Task Group, *Project Mercury [Quarterly] Status Report No. 6 for Period Ending April 30, 1960.*

Figure 35. Mercury altitude chamber in Hanger S, Cape Canaveral.

April 15

Qualification tests began on the Mercury spacecraft posigrade rocket. (See fig. 36.) The first three rocket motors subjected to these tests were successfully tested in a more stringent vibration spectrum than that required for Mercury-Atlas 1 (MA-1), the maximum dynamic reentry and maximum heat on afterbody test flight.
> NASA Space Task Group, *Project Mercury [Quarterly] Status Report No. 6 for Period Ending April 30, 1960.*

Qualification tests for the Mercury spacecraft retrorockets were started. One of the main purposes of this program was the development of a better igniter. The igniter tested was attached to the head end of the propellant grain and coated with a pyrotechnic. Based on three tests it appeared that the delayed ignition problem had been resolved. Thereafter, several other tests were run until the igniter was adjudged to be reliable.

NASA Space Task Group, *Project Mercury [Quarterly] Status Report No. 6 for Period Ending April 30, 1960.*

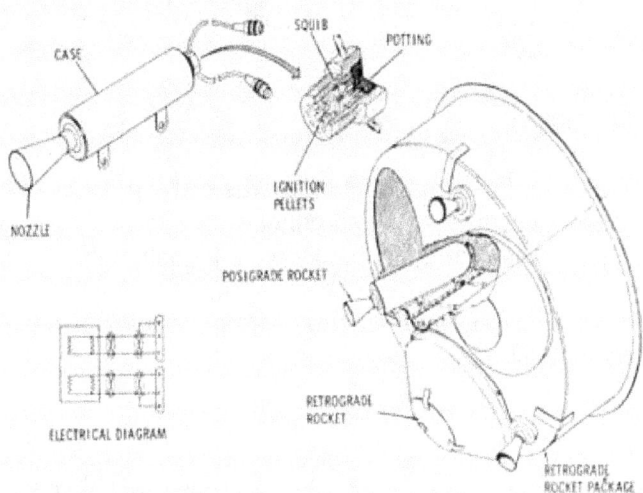

Figure 36. Posigrade rocket motors.

April 18

Fabrication of the manned environmental-control-system training spacecraft was essentially completed and a test program on the equipment was started at McDonnell. This test was completed on April 25, 1960.
NASA Space Task Group, *Project Mercury [Quarterly] Status Report No. 6 for Period Ending April 30, 1960.*

April 26

Tests were completed on the maximum altitude sensor. This component was fabricated by the Donner Scientific Company.
NASA Space Task Group, *Project Mercury [Quarterly] Status Report No. 7 for Period Ending July 31, 1960.*

April 27

Various gamma ray detectors were carried aboard Explorer XI on its orbital flight. These detectors found a directional flux of gamma radiation in space and thereby provided serious evidence against one formulation of the "steady state" cosmological theory.
Goddard Space Flight Center Chart, Satellites and Space Probe Projects as of July 1962.

April 29

Agreements, either interim or final, were concluded for all overseas Mercury tracking stations as of this date. Construction was proceeding on schedule at Cape Canaveral, Bermuda, Grand Canary Islands, the Woomera and Muchea Australian sites, and at the demonstration site on Wallops Island, Virginia. The

survey of Guaymas in Western Mexico completed that phase of the program, but the construction was yet to be accomplished.
>   NASA Space Task Group, *Project Mercury [Quarterly] Status Report No. 6 for Period Ending April 30, 1960.*

April (during the month)

Building 575, Patrick Air Force Base, Florida, was in the process of being refurbished for occupancy by NASA personnel in July 1960. This building was designated for Space Task Group use in Mercury launch, network, and data coordination.
>   NASA Space Task Group, *Project Mercury [Quarterly] Status Report No. 6 for Period Ending April 30, 1960.*

May 9

McDonnell's first production spacecraft, with its escape rocket serving as the propulsion force, was launched from Wallops Island. Designated the beach-abort test, the objectives were a performance evaluation of the escape system, the parachute and landing system, and recovery operations in an off-the-pad abort situation. The test was successful.
>   NASA Space Task Group, *Project Mercury [Quarterly] Status Report No. 7 for Period Ending July 31, 1960.*

May 12

The Space Task Group established a field representative office at the McDonnell plant in St. Louis, Missouri. A technical liaison representative, W. H. Gray, had already been assigned to the plant. A resident systems test engineer, a resident instrumentation engineer, and a team of inspectors were added to the staff.
>   Memo, Robert R. Gilruth to Space Task Group Organizational Units, subject: Organization of NASA Participation in CST at MAC, May 12, 1960.

May 14

The first production spacecraft, used in the beach-abort test, was returned to the McDonnell plant for an integrity test.
>   NASA Space Task Group, *Project Mercury [Quarterly] Status Report No. 7 for Period Ending July 31, 1960.*

May 15

Qualification tests for the Mercury spacecraft explosive egress hatch were completed.
>   NASA Space Task Group, *Project Mercury [Quarterly] Status Report No. 7 for Period Ending July 31, 1960.*

May 23

Spacecraft No. 4 (production number), after being instrumented and prepared by the Space Task Group and the Langley Research Center for flight tests, was delivered to Cape Canaveral for the first Mercury-Atlas mission (MA-1).
>Data supplied by Ken Vogel, Mercury Project Office, Manned Spacecraft Center.

May (during the month)

Training classes started for 30 physicians who had been selected by the Department of Defense to serve as medical monitors in support of Project Mercury operations. These personnel received a 2-week indoctrination program. The first week was spent at Cape Canaveral where they were briefed on the medical aspects of missile operations. The second week was spent at Space Task Group for a series of lectures and demonstrations on spacecraft systems, astronaut medical histories, and monitoring stations. This was followed by practice training sessions.
> NASA Space Task Group, *Project Mercury [Quarterly] Status Report No. 6 for Period Ending April 30, 1960.*

Production of the manned space flight configuration of the Mercury pressure suit was started. The astronauts and medical personnel who had tested the developmental suits received in November 1959 recommended a number of changes to increase the physical mobility of the astronaut before the production effort began. (See fig. 37.) Evaluation of the test suits with the suggested modifications indicated that the mobility and suit-spacecraft compatibility had been greatly enhanced. The stretching which once had been a problem area had been significantly decreased.

> NASA Space Task Group, *Project Mercury [Quarterly] Status Report No. 7 for Period Ending July 31, 1960.*

McDonnell delivered the flight-pressurized couches to be used in the animal phase of the Mercury flight test program. According to test results, the couches appeared to be satisfactory, with the exception of a slight sealing problem. McDonnell was attempting to resolve this problem.

> NASA Space Task Group, *Project Mercury [Quarterly] Status Report No. 7 for Period Ending July 31, 1960.*

May-July

During this period, two McDonnell Procedures Trainers were delivered to NASA. Number 1, delivered on May 4, 1960, was used for astronaut training in the management of the spacecraft systems at Langley Field and Number 2, delivered on July 5, 1960, was installed at Cape Canaveral, also for space flight preparations. The trainer at Langley Field, along with other equipment, later designated flight simulator, was moved in 1962 to Houston, Texas, location of the Manned Spacecraft Center, the successor to the Space Task Group.

    NASA Space Task Group, *Project Mercury [Quarterly] Status Report No. 7 for Period Ending July 31, 1960.*

Figure 37. Pressure suit worn by Alan Shepard on first manned suborbital space flight.

June 2

In considering the possible meteoroid damage to the Mercury spacecraft in orbital flight, it was concluded by the Space Task Group that damage likelihood was small even during periods of meteor showers. However, it was recommended that Mercury missions not be scheduled during forecasted shower periods.

    Memo, Benjamine J. Garland to Project Director, Space Task Group, subject: Possible Meteoroid Damage to Mercury Spacecraft, June 2, 1960.

June 3

As of this date, the funding status of Contract NAS 5-59, Mercury spacecraft, was $75,565,196.

    Memo, Glenn F. Bailey to Space Task Group Budget Officer, subject: Contract NAS 5-59 - Status of Funding, June 3, 1960.

June 9

The United States Weather Bureau estimated that it would require $50,000 during fiscal year 1961 in support of Project Mercury. Bureau responsibilities included weather forecasting for Mercury launching and recovery activities, climatological studies along the area of the Mercury ground track, and environmental studies of specified areas. With reference to the last item, a study was completed in early August 1960 of annual conditions along the Atlantic Missile Range including wind velocity, visibility and cloud coverage.
> Letter, U.S. Department of Commerce, Weather Bureau, to Dr. T. Keith Glennan, (no subject), June 9, 1960; Memo, Donald C. Cheatham to Associate Director of Project Mercury, subject: Weather in the Cape Canaveral Area, August 11, 1960.

June 18

Atlas launch vehicle 50-D was delivered for the first Mercury-Atlas mission (MA-1).
> Data supplied by Ken Vogel, Mercury Project Office, MSC.

June 20

Tests were completed on the Mercury spacecraft horizon scanner. A sandblast technique was employed in these tests, and measurements revealed that transmissibility was reduced in direct proportion to the area sand blasted. Tests covered 25, 50, and 75 percent of a germanium specimen.
> NASA Space Task Group, *Project Mercury [Quarterly] Status Report No. 7 for Period Ending July 31, 1960.*

Manned tests of the Mercury environmental control system began. (See fig. 38.) The subjects were clothed in pressure suits and subjected to postlanding conditions for 12 hours without serious physiological effects. The purpose of this test was to evaluate human tolerance, and the results indicated that no modification to the system were necessary. However, the postlanding ventilation conditions would continue to be monitored and requirements for any modifications would be evaluated.

> NASA Space Task Group, *Project Mercury [Quarterly] Status Report No. 7 for Period Ending July 31, 1960.*

June 27

As a complement to the Mercury spacecraft reliability program, a decision was made that one production spacecraft would be withdrawn from the operational program for extensive testing. The test environment would involve vacuum, heat, and vibration conditions. This test series was later designated "Project Orbit."
> Notes on Manned Space Flight Management Meeting - NASA Headquarters, June 27-28, 1960.

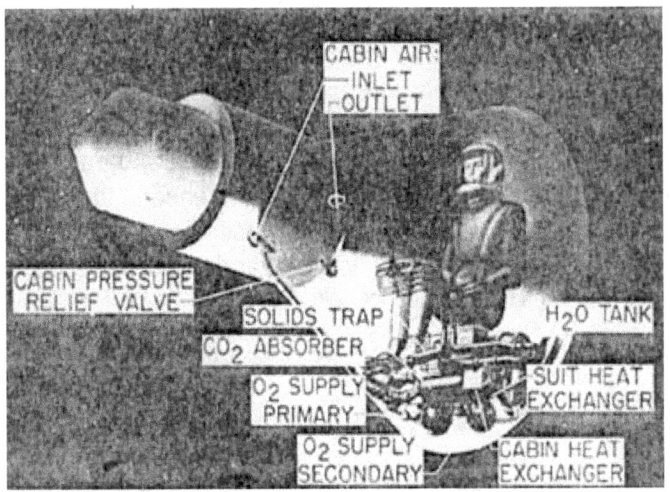
Figure 38. Mercury environmental control system.

June 30

Spacecraft No. 2 was delivered to the Marshall Space Flight Center, Huntsville, Alabama, for compatibility tests with the Redstone launch vehicle, and was shipped to Cape Canaveral on July 23, 1960.
> NASA Space Task Group, *Project Mercury [Quarterly] Status Report No. 7 for Period Ending July 31, 1960*; Data supplied by Ken Vogel, Mercury Project Office, MSC.

June (during the month)

McDonnell delivered a flight-monitoring trailer to the Space Task Group. This trailer was used at Cape Canaveral to house equipment which provided real-time telemetry read-outs during Mercury-Redstone flights.
> NASA Space Task Group, *Project Mercury [Quarterly] Status Report Nos. 6 and 7, for Period Ending April 30, 1960 and July 31, 1960.*

In the overall NASA space program, Project Mercury was the only program which included a recovery capability. For this reason, Space Task Group officials felt there were a number of experiments in the science and bioscience fields that could be placed aboard Mercury spacecraft during mission flights. An example of such experiments would be an ultra-violet camera which would provide data to assist in the design and development of an orbiting astronautical observatory; another might be bio-specimens. Obviously, decisions in experiment selections would have to be made to prevent any dilution of the primary Mercury mission.

> Notes on NASA Headquarters Manned Space Flight Management Meeting, June 27-28, 1960.

July 7

A reporting plan for Mercury-Atlas and Mercury-Redstone missions was issued. This document was amended on February 17, 1961, and April 10, 1961.
> Memo, Robert R. Gilruth to Space Task Group Division Chiefs Branch and Section Heads, subject: Reporting Plan for Mercury-Atlas and Mercury-Redstone Flight Tests, July 7, 1960.

The first meeting of the Mercury Network Coordination Committee was held at Cape Canaveral for the purpose of initiating action on existing problem areas. Subjects under review included operational procedures, range readiness, and other items associated with network operation during a mission.

> Minutes of Meeting, subject: Mercury Network Coordination Committee, July 21, 1960.

July 9

Major General Leighton I. Davis was appointed Department of Defense representative for Project Mercury support, replacing Major General Donald N. Yates.
> Information supplied by General Davis' Office, April 1963.

July 12

Beginning on this date, the astronauts underwent a five and one half day course in "desert survival" training at the Air Training Command Survival School, Stead Air Force Base, Nevada. The possibility of an arid-area landing was remote but did exist. So this training was accomplished to supply the astronaut with the confidence and ability to survive desert conditions until recovery. The course consisted of one and one half days of academics, one day of field demonstrations, and three days of isolated remote-site training. Survival equipment normally installed in the Mercury spacecraft was used to provide the most realistic conditions.
> NASA Space Task Group, *Project Mercury [Quarterly] Status Report No. 7 for Period Ending July 31, 1960.*

July 14

Personnel strength in support of Project Mercury was 543. This included 419 assigned to the Space Task Group, and 124 personnel from the Langley Research Center.
> Memo, Paul E. Purser to Charles J. Donlan, Associate Director of Space Task Group, subject: Study of Space Task Group Personnel Needs for FY 1961, July 14, 1960.

July 23

Mercury spacecraft No. 2 was delivered to Cape Canaveral for the Mercury-Redstone 1-A (MR-1A) mission.
> Data supplied by Ken Vogel, Mercury Project Office, MSC.

July 27

Mercury launch site recovery forces exercised in recovery operations following simulated spacecraft landings off Cape Canaveral. Coordination and control of the recovery forces were rated highly satisfactory.
> NASA Space Task Group, *Project Mercury [Quarterly] Status Report No. 8 for Period Ending October 31, 1960.*

July 29

Mercury spacecraft No. 3 was delivered to Langley Field for a noise and vibration test.
> NASA Space Task Group, *Project Mercury [Quarterly] Status Report No. 7 for Period Ending July 31, 1960.*

Mercury-Atlas (MA-1) was launched from Cape Canaveral with mission objectives being to check the integrity of the spacecraft structure and afterbody shingles for a reentry associated with a critical abort and to evaluate the open-loop performance of the Atlas abort-sensing instrumentation system. (See fig. 39.) The spacecraft contained no escape system and no test subject. Standard posigrade rockets were used to separate the spacecraft from the Atlas, but the retrorockets were dummies. About 59 seconds after launch, the flight was terminated because of a launch vehicle and adapter structural failure. The spacecraft was destroyed upon impact with the water because the recovery system was not designed to actuate under the imposed flight conditions. Later most of the spacecraft, the booster engines, and the liquid oxygen vent valve were recovered from the ocean floor. Since none of the primary flight objectives was achieved, Mercury-Atlas 2 (MA-2) was planned to fulfill the mission.

> Memo, George Low to NASA Administrator, subject: Mercury-Atlas 1, Post Launch Information, July 29, 1960; NASA Space Task Group, *Project Mercury [Quarterly] Status Report No. 7 for Period Ending July 31, 1960.*

July (during the month)

Manufacture of the mobile-pad egress tower (cherry picker) was completed (fig. 40), and the vehicle was delivered to Cape Canaveral on October 24, 1960.
> NASA Space Task Group, *Project Mercury [Quarterly] Status Report No. 8 for Period Ending October 31, 1960.*

August 1

Marshall Space Flight Center published the "Final Standard Trajectory for MR-1 (Mercury-Redstone)."
> Report, MNN-M-AERO-2-60, Aug. 1, 1960.

August 3

Redstone launch vehicle No. 1 was delivered to Cape Canaveral for the MR-1 (Mercury-Redstone).
> Data supplied by Ken Vogel, Mercury Project Office, MSC.

Figure 39. Mercury-Atlas 1.

Figure 40. Mobile pad egress tower (Cherry picker).

August 10

The Wright Air Development Center requested that NASA Headquarters provide the Center with pertinent working papers and reports on Project Mercury, especially on human factor aspects, for possible application in the X-20 Dyna Soar program.
> Letter, Wright Air Development Center to NASA Hqs., subject: Project Mercury Technical Data for Use in Dyna Soar Programing, Aug. 10, 1960.

August 11

Representatives of NASA, McDonnell, Ballistic Missile Division, Space Technology Laboratories, and Convair met at Cape Canaveral and later at Convair Astronautics (Aug. 30, 1960) to discuss the Mercury-Atlas 1 (MA-1) mission malfunction. James A. Chamberlin of the Space Task Group was appointed chairman of a joint committee to resolve the problems and to provide a solution prior to the Mercury-Atlas 2 (MA-2) mission. Work accomplished at this meeting is as follows: A complete analysis of Mercury-Atlas 1 flight data and correlation of the data with data of all previous Atlas flights; a special dynamic load analysis; study of vibration tests of spacecraft, adapter, and the Atlas upper tank section; and review of wind tunnel studies of buffeting loads on spacecraft, adapter, and the Atlas upper tank sections.
> Report, subject: Atlas Mercury Failure, Examination of Failed Parts by J. A. Kies, Aug. 30, 1960; Trip Report by Andre J. Meyer, Jr., subject: Mercury-Atlas Failure, Aug. 30, 1960; Memo, Warren J. North to NASA Administrator, subject: Analysis of MA-1 Malfunction, Aug. 22, 1960.

The Mercury spacecraft landing system qualification test program was completed. The entire qualification testing program consisted of 56 airdrops of full-scale engineering models of the Mercury spacecraft from C-130 aircraft at various altitudes up to 30,000 feet and from helicopters at low altitudes to simulate off-the-pad abort conditions. This test program, under contract to Northrop, had spanned one and one half years.

> NASA Space Task Group, *Project Mercury [Quarterly] Status Report No. 8 for Period Ending October 31, 1960.*

August 12

Weather Bureau fund estimates for Fiscal Year 1961 for support of Project Mercury were adjusted to $180,000, but in April 1961, the Bureau Director stated he believed that actual costs would not exceed $150,000.
> Letter, U.S. Department of Commerce, Weather Bureau, to Robert R. Gilruth (no subject), August 12, 1960 and April 18, 1961.

August 16-18

At the design engineering inspection of spacecraft No. 7, a number of requests for changes in the control panel area were made by the astronauts to facilitate pilot operation. Later, meeting procedures for design engineering inspections were standardized and conducted by a permanent team at appropriate intervals.
> NASA Space Task Group, *Project Mercury [Quarterly] Status Report No. 8 for Period Ending October 31, 1960.*

August 26

Coordination effectiveness among organizations directly involved in the Mercury development and test program was reviewed by the Space Task Group at the request of NASA Headquarters. Conclusions were that the interchange of information had been excellent. The coordination panel meetings were cited as a fine medium for information exchange. The Mercury-Atlas Coordination Panel first met on February 19, 1959, and by the date of the review, a total of 29 days had been spent in these meetings. Interchange of visits had started even before the cited February date and had been continued with good results.
> Letter, Space Task Group to NASA Headquarters, subject: Project Mercury Coordination between NASA-MAC and BMD-STL-Convair, Aug. 26, 1960.

August (during the month)

Astronaut side-hatch-egress training was completed with no difficulties encountered. The astronauts later received refresher training prior to mission flights. In fact, during the refresher phases, better procedures were developed. An example was the helicopter mode in which a line was attached to the top of the spacecraft and the spacecraft was partially raised by the helicopter. Then, the astronaut emerged from the side egress hatch and was raised by a second line to the helicopter. (See fig. 41)

> NASA Space Task Group, *Project Mercury [Quarterly] Status Report No. 8 for Period Ending October 31, 1960.*

The astronauts were briefed on the Tiros weather satellite project as a means providing them with information that could be used to recognize and report on weather phenomena during orbital flight.

> NASA Space Task Group, *Project Mercury [Quarterly] Status Report No. 8 for Period Ending October 31, 1960.*

The first phase of the program in which boilerplate spacecraft with impact skirts were dropped by helicopters on water and land surfaces was completed. These tests were performed to investigate spacecraft dynamics, effects of parachute restraint and release time on spacecraft dynamics, and to determine maximum landing decelerations. During the drops into the water spacecraft water stability was shown to be unacceptable, because a portion of the spacecraft cylindrical section remained under water. McDonnell immediately investigated this problem and performed such experiments as redistribution of weight to obtain center-of-gravity positions which were acceptable but yet provided satisfactory flotation characteristics. Space Task Group was investigating the possibility of extending the heat shield from the remainder of the spacecraft and thereby creating a greater stabilizing moment. Results from the drops on land appeared to be acceptable because of the relatively low decelerations and the overall low probability of a landing on land.

> NASA Space Task Group, *Project Mercury [Quarterly] Status Report No. 8 for Period Ending October 31, 1960.*

Tests conducted by Space Task Group personnel proved fluorescein green dye dispersed from a floating disc-shaped canister was superior to other products for this phase of Mercury recovery operations. This material had been used previously, but it had been briefly discarded in favor of an aluminium-colored dye. However, the new type proved to be unsatisfactory and the use of the green dye marker was resumed.

> NASA Space Task Group, *Project Mercury [Quarterly] Status Report No. 8 for Period Ending October 31, 1960.*

Figure 41. Mercury spacecraft and astronaut Shepard being recovered by Marine Corps helicopter.

August 1960 to February 1961

Because of the failure of the Big Joe Atlas test flight and the Mercury-Atlas 1 (MA-1) flight to attain all its mission objectives, the overall Mercury-Atlas program underwent an exhaustive review. In the Big Joe firing, velocity and range had been considerably below nominal values because the launch vehicle had failed to stage, and spacecraft separation had been delayed because of recontact. In the Mercury-Atlas 1 flight, launch vehicle performance was normal until about 57.6 seconds of flight, and the launch vehicle was destroyed at 59 seconds. Neither flight had sufficient instrumentation to pinpoint the exact cause of the failures; therefore, an extensive evaluation and test program was initiated. Meetings on these matters began immediately among the interested parties to coordinate findings and recommendations for solutions (for instance, Aug. 9 -

summary evaluation of Mercury-Atlas 1 data at Los Angeles; Aug. 11 - evaluation summary meeting at the Atlantic Missile Range; Aug. 22 - Investigation Panel meeting at McDonnell; Sept. 9 - Investigation Panel meeting at Convair Astronautics; Sept. 14 - management meeting at Atlantic Missile Range; Sept. 26 - Instrumentation and Wind Tunnel Test Conference at Space Task Group; Oct. 3-8 - Vibration Tests at McDonnell; Oct. 3-8 - wind tunnel tests at the Arnold Engineering Development Center; and Nov. 16 - test program summary at Space Task Group. During the course of these meetings and tests, it was the considered opinion of Space Task Group and other interested parties that the trouble had developed in the spacecraft interface area. One of the tests involved stiffening the adapter rings, and later tests showed that this solution was quite satisfactory. Tests also showed there were some moderately high stresses in the launch vehicle near a welded joint just aft of the adapter, and this area was strengthened by adding a band stiffener, which proved to be satisfactory. It was also decided for the upcoming Mercury-Atlas 2 (MA-2) mission that additional instrumentation would be integrated with the spacecraft and launch vehicle in order to define loads on the vehicle in the interface area, to measure pressure on and in the adapter, and to measure any undue responses in this area. Still another decision was that the Atlas launch vehicle, commencing with Mercury-Atlas 3 (MA-3) would be a "thick-skin" configuration. These findings and recommendations were presented to a NASA/Air Force *ad hoc* group on February 13 through 17, 1961, commonly known as the Rhode (NASA)-Worthman (Air Force) committee. The committee studied the Space Task Group proposals for the Atlas launch vehicle and adapter modifications and approved the test findings and the contemplated action.

>Notes maintained by Paul E. Purser, Special Assistant to Director, Manned Spacecraft Center, covering cited period.

September 1

The Space Task Group drafted and forwarded to McDonnell the specification requirements for spacecraft on-board data system instrumentation tests. McDonnell was to demonstrate the satisfactory performance of all space communication and instrumentation systems.

>Letter, Space Task Group to Walter Burke, McDonnell, subject: Contract NAS 5-59; Proposed On-Board Data System Instrumentation Tests, Sept. 7, 1960, with inclosures.

Mercury spacecraft No. 6 was delivered to Cape Canaveral for the Mercury-Atlas 2 (MA-2) unmanned mission intended to gain data on maximum dynamic pressure and maximum heat on the spacecraft afterbody.

>Data supplied by Ken Vogel, Mercury Project Office, MSC.

September 3

Aircraft telemetry requirements were deleted from the Mercury-Atlas 3 (MA-3) and Mercury-Atlas 4 (MA-4) missions, as the spacecraft had been modified to provide telemetry transmissions from the point of main parachute deployment to landing.

        Letter, Space Task Group to Air Force Missile Test Center, subject: T/M Aircraft to Support MA-3 and MA-4 Operations, Dec. 8, 1960.

September 9

McDonnell forwarded its plans to the Space Task Group for the spacecaft systems tests and Cape Canaveral checkout plans for spacecraft Nos. 5 and 7. Later, spacecraft No. 7 was the first to undergo this type of test.

        Informal Memo, J. F. Yardley and G. M. Preston, Space Task Group, to (Conference) Attendees, subject: Summary of Conclusion Reached Regarding CST Plans and Cape Checkout Plans for Capsules 5 and 7, Sept. 9, 1960.

September 12

"Flight Test Evaluation Report, Missile 50-D", Report No. AE 60-0323, was published. The launch vehicle was used in the unsuccessful Mercury-Atlas 1 (MA-1) reentry test mission.

        Source as cited in text.

September 19

The format of subject matter coverage for the first Mercury-Redstone postlaunch (MR-1) report was issued. This report, covering a full range of topics related to the mission, was to be submitted within 5 days after the launch.

        Memo, Robert R. Gilruth, Director of Project Mercury to those concerned, subject: MR-1 Postlaunch Report, Sept. 19, 1960, with inclosures.

September 20

The Atlas launch vehicle 67-D was delivered to Cape Canaveral for the Mercury-Atlas 2 (MA-2) reentry test mission.

        Data supplied by Ken Vogel, Mercury Project Office, MSC.

September 21

Because of poor tower separation of the production spacecraft in the off-the-beach abort test at Wallops Island, NASA personnel at Langley started a series of jettison rocket tests. It was found that rocket performance had been only about 42 percent of the desired level, and experiments were started to raise thrust effectiveness. Measures taken included canting the motor, adding a cone to the blast shield, and, in one instance, deleting the blast shield. Space Task Group personnel advised McDonnell that plans were made to test a redesigned jettison rocket nozzle, consisting of three nozzles spaced 120 degrees apart and canted at a 30 degree angle to the rocket centerline. (See fig. 42.) The three-nozzle effect, which produced the desired results, was another NASA engineering contribution.

NASA Space Task Group, *Project Mercury [Quarterly] Status Report No. 8 for Period Ending October 31, 1960.*

The astronauts received weightless training in a modified C-135 jet aircraft. This was the third type of aircraft used by the astronauts in such training. The previously mentioned F-100 provided a weightless period of some 40 to 50 seconds; the C-131, 15 seconds; and the C-135, 30 seconds. During the C-135 flights, the astronauts were checked for changes in normal speech and their ability to control a tracking problem while undergoing moderate g-loads prior to entering the weightless periods.

NASA Space Task Group, *Project Mercury [Quarterly] Status Report No. 8 for Period Ending October 31, 1960.*

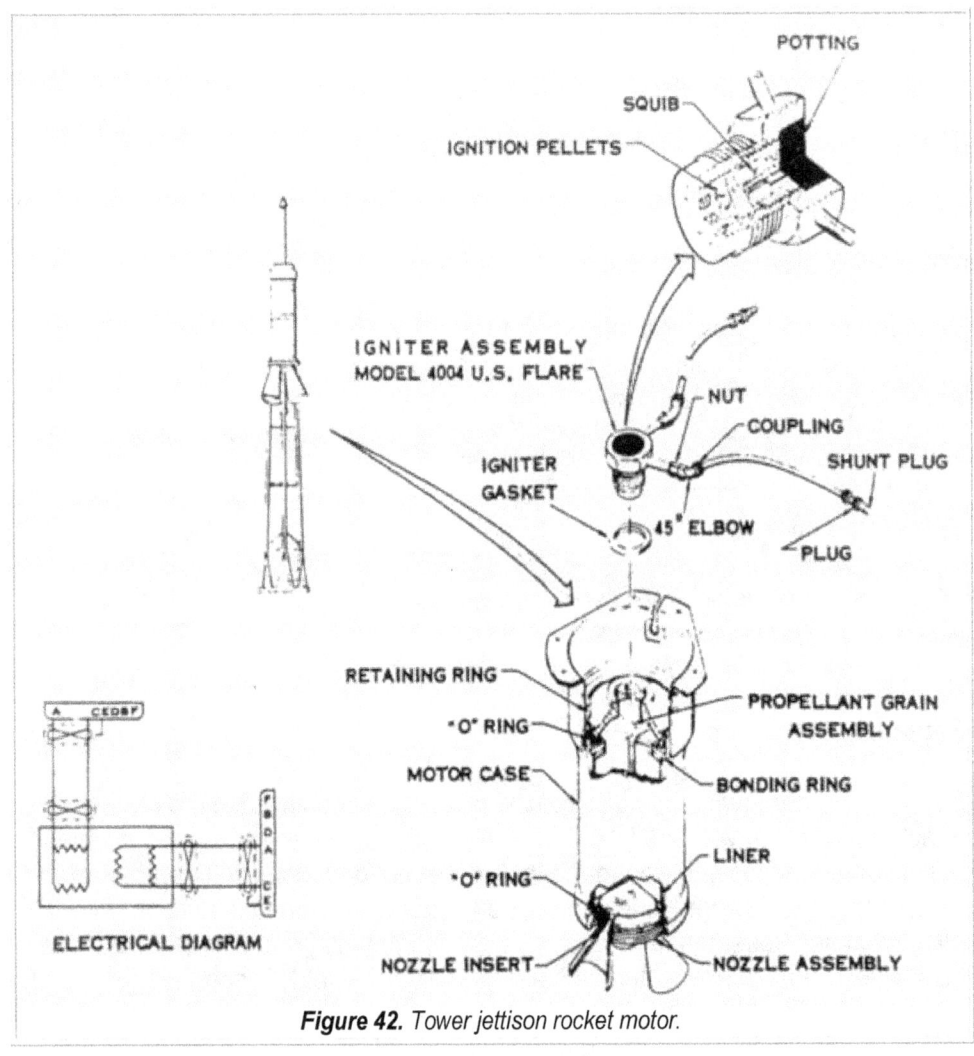

Figure 42. *Tower jettison rocket motor.*

September 26

The roll-out inspection of Atlas launch vehicle 77-D was conducted at Convair-Astronautics. This launch vehicle was allocated for the Mercury-Atlas 3 (MA-3) mission, but was later canceled and Atlas booster 100-D was used instead.
> NASA Space Task Group, *Project Mercury [Quarterly] Status Report No. 8 for Period Ending October 31, 1960*; Data supplied by Ken Vogel, Mercury Project Office, MSC.

September 27

Mercury spacecraft No. 3, initially delivered to Langley on July 29, 1959, for a noise and vibration test, was erected at the Wallops Island launch site for the Little Joe 5 (LJ-5).
> Data supplied by Ken Vogel, Mercury Project Office, MSC.

September 30

Mercury spacecraft No. 5 was delivered to the Marshall Space Flight Center for booster compatibility checks, and was shipped to Cape Canaveral on October 11, 1960, for the Mercury-Redstone 2 (MR-2) ballistic-primate (Ham) mission.
> Data supplied by Ken Vogel, Mercury Project Office, MSC.

September (during the month)

Flight-type pressure suits were received from the B. F. Goodrich Company and were immediately used on the human centrifuge to assist in determining final adjustments that were necessary in preparation for manned space flight.
> NASA Space Task Group, *Project Mercury [Quarterly] Status Report No. 8 for Period Ending October 31, 1960*.

October 3-21

The third centrifuge training program was conducted for the astronauts at the Aviation Medical Acceleration Laboratory. This was considered the final major centrifuge training preparation for the first manned Mercury-Redstone flight. No difficulties were encountered; a decided improvement in the performance of 3-axis hand-controller tasks by the astronauts was noted. The Mercury-Redstone 3 (MR-3) flight activities were adhered to as closely as possible - actual spacecraft couches were used, a production hand-controller assembly was installed, the latest model pressure suits were worn, and the environmental control system was equipped with a freon coolant. Failures in spacecraft sequencing were introduced which required the astronaut to initiate an appropriate manual override.
> NASA Space Task Group, *Project Mercury [Quarterly] Status Report No. 8 for Period Ending October 31, 1960*.

October 13-14

DESFLOTFOUR personnel, designated previously by the Department of Defense to provide recovery support for Project Mercury, conducted a communications exercise in the recovery room of Mercury Control Center. This was the first time these communication facilities had been used since the installation of the equipment. During the exercise, voice and continuous-wave communications were established with two destroyers 120 miles at sea. The purpose of this successful exercise was to acquaint personnel with equipment layout and communication procedures.
>NASA Space Task Group, *Project Mercury [Quarterly] Status Report No. 8 for Period Ending October 31, 1960.*

October 17

A Project Mercury weather support group was established in the Office of Meteorological Research of the United States Weather Bureau at the request of NASA.
>Emme, *Aeronautics and Astronautics: 1915-1960*, p. 129.

James Carter of the Marshall Space Flight Center submitted a study on "Crew Support Equipment." This type of equipment was defined as that which is not an integral part of or attached to a space vehicle or space station. Specific equipment categories discussed in the report included personal safety, recovery, survival, food supplies, portable respiratory devices, and hand tools.

>Report, MTP-M-FPO-1-60, Marshall Space Flight Center, subject: Crew Support Equipment, Oct. 17, 1960.

October 18

The spacecraft checkout facility at Marshall Space Flight Center was transferred to Cape Canaveral.
>Memo, Warren J. North to NASA Director of Space Flight Programs, subject: Significant Items Within Mercury Program Management Chart, Oct. 18, 1960.

Mission rules for Mercury-Redstone 1 (MR-1) were issued. A revision was published on Nov. 1, 1960.

>Project Mercury, MCC MR-1, subject: Project Mercury Control Center Operations, and Flight Control Procedures and Countdown, Oct. 18, 1960.

October 31

Space Task Group officials presented the status of qualification and reliability activities for Project Mercury to Dr. T. Keith Glennan, NASA Administrator.
>Memo, George Low to NASA Director of Program and Analysis Control, subject: Project Mercury Briefing, Oct. 31, 1960.

November 1

The Goddard Space Flight Center computing and communications center became operational. Goddard's mission was to serve as a communications center, and two IBM 7090 computers, operating in parallel, would compute the smoothed exact position at all times during the flight, predict future spacecraft positions, and shift the coordinates to provide acquisition information for all observation sites. (See fig. 43.) In addition, Goddard calculated certain quantities needed for display purposes at Cape Canaveral, Florida. The importance of the Goddard computers was graphically demonstrated when they predicted the amount of overshoot within seconds after landing during the Mercury-Atlas 7 (MA-7, Carpenter) mission. This action significantly reduced the time to find and recover the astronaut.
> NASA Space Task Group, *Project Mercury [Quarterly] Status Report No. 9 for Period Ending January 31, 1961.*

Figure 43. Computers used in Mercury orbital track at Goddard Space Flight Center.

November 8

Little Joe 5 (LJ-5), the first of the series with a McDonnell production spacecraft, was launched from Wallops Island to check the spacecraft in an abort simulating the most severe launch conditions. The launch was normal until 15.4 seconds after lift-off, at which time the escape rocket motor was prematurely ignited. The spacecraft did not detach from the launch vehicle until impact and was destroyed. Failure to attain mission objectives was attributed to several possible causes. One of these was failure of the spacecraft-to-adpater clamp-ring limit switches. Another possibility was failure of the escape tower clamp-ring limit switches. And the third was improper rigging of the limit switches in either of those locations so that vibration or deflection could have caused switch closure. Since the test objectives were not met, a repeat of the mission was planned.
> Memo, George Low to NASA Administrator, subject: Report On Little Joe No. 5 and Mercury-Redstone 1, Nov. 10, 1960; NASA Space Task Group, *Project Mercury [Quarterly] Status Report No. 9 for Period Ending January 31, 1961.*

November 13

System checkout tests were completed on spacecraft No. 7. In the opinion of McDonnell, the results demonstrated that this spacecraft was adequate for a manned mission.
> Memo, James T. Rose, MR-3 Assistant Project Engineer, to W. H. Gray, Space Task Group Liaison Officer to McDonnell, subject: General Summary of Capsule Systems Test on Capsule No. 7, Dec. 1, 1960 with inclosures.

November 16

A meeting was held at Langley Field by NASA personnel to discuss the results of test programs which had been conducted. Of particular interest was the establishment of the causes for the failure of the Mercury-Atlas 1 (MA-1) mission and to determine the status of readiness or the Mercury-Atlas 2 (MA-2) mission. (See August 1960 to February 1961 entry.)
> Minutes of Meeting, MA-2, subject: Summary of Test Programs and Recommendations for MA-2 Launch, Nov. 16, 1960.

November 17

The Space Task Group requested that McDonnell submit a proposal for conducting a test to determine the capability of an astronaut to make celestial observations through the Mercury spacecraft observation window.
> Space Task Group Message PAM-0027, NASA Space Task Group to Walter F. Burke, Vice President, McDonnell Aircraft Corporation, subject: Project Mercury, Contract NAS 5-59, Nov. 17, 1960.

November 18

The "Standard Procedures Mercury Control Center for Flight Control and Overall Options" was published.
> Project Mercury MCC SP, Nov. 18, 1960.

Spacecraft No. 8 was delivered to Cape Canaveral for the Mercury-Atlas 3 (MA-3) unmanned orbital mission.

> Data supplied by Ken Vogel, Mercury Project Office, MSC.

November 21

An attempt was made to launch Mercury-Redstone 1 (MR-1) from Cape Canaveral. This unmanned mission was unsuccessful because premature cut-off of the launch vehicle engines activated the emergency escape system when the vehicle was only about 1 inch off the pad. Engine cut-off was caused by premature loss of electrical ground power to the booster. The launch vehicle settled back on the pad with only slight damage. Since the spacecraft received a

cut-off signal, the escape tower and recovery sequence was initiated. The undamaged spacecraft was recovered for reuse.
> Memo, George Low to NASA Administrator, subject: Attempted Launching of MR-1, Nov. 21, 1960.

November 21-30

Phase II of the helicopter spacecraft airdrop program was completed. One of the objectives of these tests was to drop a spacecraft during wind conditions of 18 knots, and this phase was successful. Secondary objectives of the program were to investigate spacecraft dynamics and water stability. Both spacecraft flotation and righting characteristics were found to be acceptable.
> NASA Space Task Group, *Project Mercury [Quarterly] Status Report No. 9 for Period Ending January 31, 1961.*

November-December

During the Mercury-Redstone 1 (MR-1) and Mercury-Redstone 1A (MR-1A) launches, the complete Mercury Control Center staff operated for the first time.
> NASA Space Task Group, *Project Mercury [Quarterly] Status Report No. 9 for Period Ending January 31, 1961.*

December 1

A 16 and one half foot recovery whip antenna replaced the balloon-borne system on the Mercury spacecraft. (See fig. 44.)
> NASA Space Task Group, *Project Mercury [Quarterly] Status Report No. 9 for Period Ending January 31, 1961.*

McDonnell completed the fabrication of the first spacecraft orbital timing device, and qualification tests for this component were started immediately.

> NASA Space Task Group, *Project Mercury [Quarterly] Status Report No. 9 for Period Ending January 31, 1961.*

December 2

Spacecraft weight and balance values for the Mercury-Redstone 2 (MR-2) mission were forwarded by the Space Task Group to the Marshall Space Flight Center.
> Message, NASA Space Task Group to Marshall Space Flight Center, subject: Calculated Weight and Balance Values for Capsule 5, MR-2, Dec. 2, 1960.

December 3

Redstone launch vehicle No. 3 was shipped to Cape Canaveral for the Mercury-Redstone 1A (MR-1A) mission.
> Data supplied by Ken Vogel, Mercury Project Office, MSC.

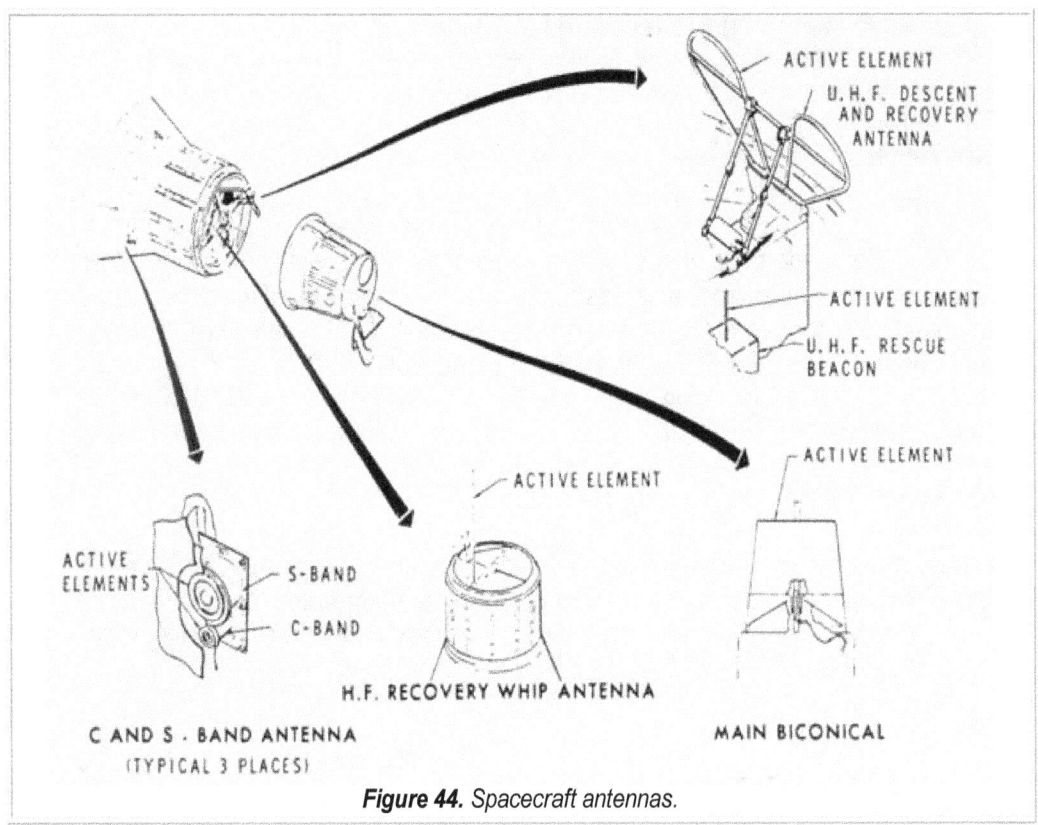
Figure 44. Spacecraft antennas.

December 9

Spacecraft No. 7 was delivered to Cape Canaveral for the Mercury-Redstone 3 (MR-3) manned ballistic mission (Shepard).
  Data supplied by Ken Vogel, Mercury Project Office, MSC.

December 14

A contract with the Waltham Precision Instrument Company for the development of a satellite clock was canceled. Technical difficulties were encountered in the manufacturing of the device, previously scheduled for delivery in August 1960, and there was little assurance that these problems could be resolved in time for the clock to be used in any of the Mercury flights. McDonnell fabricated an orbital timing device, which proved to be very satisfactory.
  NASA Space Task Group, *Project Mercury [Quarterly] Status Report No. 9 for Period Ending January 31, 1961.*

December 19

Mercury-Redstone 1A (MR-1A) was launched from Cape Canaveral in a repeat of the November 21, 1960, mission and was completely successful. This was the third attempt to accomplish the objectives established for this flight. The first attempt on November 7, 1960, was canceled as a result of a helium leak in the spacecraft reaction control system relief valve, and on November 21, 1960, the mission could not be completed because of premature cut-off of the launch vehicle engines. Objectives of the MR-1A flight were to qualify the spacecraft for space flight and to qualify the flight system for a primate flight scheduled shortly thereafter. Close attention was given to the spacecraft-launch vehicle combination as it went through the various flight sequences: powered flight; acceleration and deceleration; performance of the posigrade rockets; performance of the recovery system; performance of the launch, tracking, and recovery phases of the operation; other events of the flight including retrorocket operation in a space environment; and operation of instrumentation. Except that the launch vehicle cut-off velocity was slightly higher than normal, all flight sequences were satisfactory; tower separation, spacecraft separation, spacecraft turnaround, retrofire, retropackage jettison, and landing system operation occurred or were controlled as planned. The spacecraft reached a maximum altitude of 130.68 statute miles, a range of 234.8 statute miles, and a speed of 4,909.1 miles per hour. Fifteen minutes after landing in the Atlantic Ocean, the recovery helicopter picked up the spacecraft to complete the successful flight mission.

> Memo, George Low to NASA Administrator, subject: Mercury-Redstone 1(A) Launching, Dec. 20, 1960; NASA Space Task Group, *Project Mercury [Quarterly] Status Report No. 9 for Period Ending January 31, 1961.*

December 20

Redstone launch vehicle No. 2 was delivered to Cape Canaveral for the Mercury-Redstone 2 (MR-2) mission (chimpanzee "Ham" flight).
> Data supplied by Ken Vogel, Mercury Project Office, MSC.

1961

January 3

The Space Task Group, charged by NASA to conduct Project Mercury and other manned space-flight programs, officially became a separate NASA field element directly under NASA Headquarters. Prior to this time, the Space Task Group was organized under the Goddard Space Flight Center and was administratively supported by the Langley Research Center. As of this date, the personnel strength of Space Task Group was 667.
> House Committee, *Aeronautical and Astronautical Events of 1961*, June 7, 1962.

January 16

The Mercury-Redstone 1A (MR-1A) postlaunch system evaluation tests were completed at Cape Canaveral. Data disclosed that the instrumentation system, communication system, and other components had operated satisfactorily during the flight mission.

NASA Space Task Group, *Project Mercury [Quarterly] Status Report No. 9 for Period Ending January 31, 1961.*

January 20

Spacecraft No. 14 was delivered to Wallops Island for the Little Joe 5A (LJ-5A) maximum dynamic pressure abort test.
   Data supplied by Ken Vogel, Mercury Project Office, MSC.

January 31

The estimated cost of NASA Order HS-36, Atlas launch vehicles, was $51,504,000, of which, definitive documents in the amount of $43,671,000 had been processed as of the cited date. NASA Order HS-44 for Redstone launch vehicles was $14,918,182 and $12,534,182 had been processed. On contract NAS 5-59, Mercury spacecraft, costs were $79,245,952, and approximately $9.5 million of this figure was classed as "Undefinitized Obligations."
   Memo, NASA-STG Procurement and Supply Office to NASA Hqs., subject: Monthly Status Report, February 10, 1961.

Mercury-Redstone 2 (MR-2) was launched from Cape Canaveral, with Ham, a 37-pound chimpanzee aboard the spacecraft. (See fig. 45.) During the powered phase of the flight, the thrust of the propulsion system was considerably higher than planned. In addition, the early depletion of the liquid oxygen caused a signal that separated the spacecraft from the launch vehicle a few seconds before planned. The over-acceleration of the launch vehicle coupled with the velocity of the escape rocket caused the spacecraft to attain a higher altitude and a longer range than planned. However spacecraft recovery was effected, although there were some leaks and the spacecraft was taking on water. Ham appeared to be in good physiological condition, but sometime later when he was shown the spacecraft it was visually apparent that he had no further interest in cooperating with the space flight program. Despite the over-acceleration factor, the flight was considered to be successful.

   Memo, Warren J. North to Franklyn W. Phillips, NASA Code A, subject: MR-2 Flight Results, February 1, 1961.

As of this date, McDonnell had expended 2,616,387 man-hours in engineering; 383,561 man-hours in tooling, and 1,538,476 man-hours in production in support of Project Mercury.

   Letter, McDonnell Aircarft Corporation to Space Task Group, subject: Contract NAS 5-59, Monthly Financial Report, Feb. 24, 1961.

Figure 45. Chimpanzee, "Ham," flown in Mercury-Redstone 2 suborbital flight.

January (during the month)

Astronaut training was centered on a close study of spacecraft systems in final preparation for manned space flight. A series of lectures was presented to the astronauts by the Operations Division of the Space Task Group in this respect.
    NASA Space Task Group, *Project Mercury [Quarterly] Status Report No. 9 for Period Ending January 31, 1961.*

February 3

The Eagle-Picher Company started a 13-week life-cycle test on the Mercury spacecraft batteries.
    NASA Space Task Group, *Project Mercury [Quarterly] Status Report No. 10 for Period Ending April 30, 1961.*

February 10

Mission rules for the Mercury-Redstone 3 (MR-3 - Shepard's flight) were published. Revisions were issued on February 27, and April 28, 1961.
    Project Mercury MCC MR-3, subject: Mercury Control Center Countdown, Flight Control and Overall Operations, Feb. 10, 1961.

Measures to be taken for hydrogen-peroxide fuel economy for the spacecraft attitude control system were studied at a coordination meeting. Items considered were orbital attitude, retroattitude hold sequence, and salvo versus ripple retrorocket firing. Astronaut Virgil Grissom reported that the salvo method had already been proven to be unsatisfactory on the Mercury procedures trainer.

> Minutes, Group III Meeting 19, subject: Project Mercury, Project Coordination Meeting, February 10, 1961.

February 15

After his nomination by the President as Administrator of NASA on January 30, 1961, James E. Webb was sworn into office, replacing T. Keith Glennan.
> House Committee, *Aeronautical and Astronautical Events of 1961*, June 7, 1962.

February 17

The Space Task Group requested that McDonnell design and install a manual bilge pump in spacecraft No. 7 to allow the removal of any seawater resulting from leakage after spacecraft impact.
> Minutes, Group II Meeting No. 23, subject: Project Mercury Coordination Meeting, March 8, 1961; Message PAM-0102, Space Task Group to McDonnell Aircraft Corporation, Feb. 17, 1961.

Information was released by NASA Headquarters that Space Task Group engineers directing Project Mercury had selected the flight trajectory for the Mercury-Atlas 2 (MA-2) mission. This trajectory was designed to provide the most severe reentry heating conditions which could be encountered on an emergency abort during an orbital flight attempt. The reentry heating rate was estimated to be 30 percent higher than a normal Mercury orbital reentry, and temperatures were predicted to be about 25 percent higher at certain locations on the afterbody of the spacecraft. In addition, the deceleration g-load was calculated to be about twice that expected for a normal reentry from orbit.

> Notes by Paul Haney, NASA Hqs., subject: Mercury Spacecraft Flight Test (MA-2), Feb. 17, 1961.

Egress hatch procedures for recovery force operations were discussed at a coordination meeting. One suggestion involved the installation of a pull-ring for activating the hatch explosive charge. Another proposal was made for a paint outline of an emergency outlet that could be cut through, if necessary.

> Minutes, Project Mercury Coordination Meeting, February 17, 1961, issued Feb. 17, 1961.

February 17-20

Spacecraft, mission, and launch vehicle flight safety rules for the Mercury-Atlas 2 (MA-2) mission were reviewed by Space Task Group personnel.
> Mercury-Atlas 2 Mission Rules (Capsule No. 6), Feb. 17-20, 1961.

February 21

Mercury-Atlas 2 (MA-2) was launched from Cape Canaveral in a test to check maximum heating and its effects during the worst reentry design conditions. The flight closely matched the desired trajectory and attained a maximum altitude of 114.04 statute miles and a range of 1,431.6 statute miles. Inspection of the spacecraft aboard the recovery ship some 55 minutes after launch (actual flight time was 17.56 minutes) indicated that test objectives were met, since the structure and heat protection elements appeared to be in excellent condition. The flight control team obtained satisfactory data; and the complete launch computing and display system, operating for the first time in a flight, performed satisfactorily.
> Memo, Warren J. North to NASA Administrator, subject: Preliminary MA-2 Flight Results, Feb. 23, 1961.

Astronauts John Glenn, Virgil Grissom, and Alan Shepard were selected by the Space Task Group to begin special training for the first manned Mercury flight.

> House Committee, *Aeronautical and Astronautical Events of 1961*, June 7, 1962.

February 23

As of this date, the Space Task Group, Convair-Astronautics, Space Technology Laboratories, McDonnell, and the Marshall Space Flight Center had completed a number of extensive studies on the subject of the safe separation of the Mercury spacecraft from the launch vehicle during an emergency. The following papers include a report of these studies: NASA Project Mercury Working Paper No. 111, "Mercury-Redstone Separation Distance ..."; NASA Project Mercury Working Paper No. 141, "Dispersion Study of Separation Distance ...for Mercury-Redstone"; and NASA Working Paper No. 152, "Determination of Mercury Escape Rocket Thrust Eccentricity ...from Mercury-Atlas Booster."
> Letter, Space Task Group to Thiokol Chemical Corporation (no subject), Feb 23, 1961, with enclosures.

February 24

Spacecraft No. 9 was delivered to Cape Canaveral for the Mercury-Atlas 5 (MA-5) orbital primate (Enos) mission.
> Data supplied by Ken Vogel, Mercury Project Office, MSC.

February 25

McDonnell conducted a successful drop test, using a boilerplate spacecraft fitted with impact skirt, straps and cables, and a beryllium heat shield. During the tests the stainless steel straps were successfully stretched to design limits. (See fig. 46.)

> Minutes, Group I Meeting 22, subject: Project Mercury Coordination Meeting of February 27, 1961.

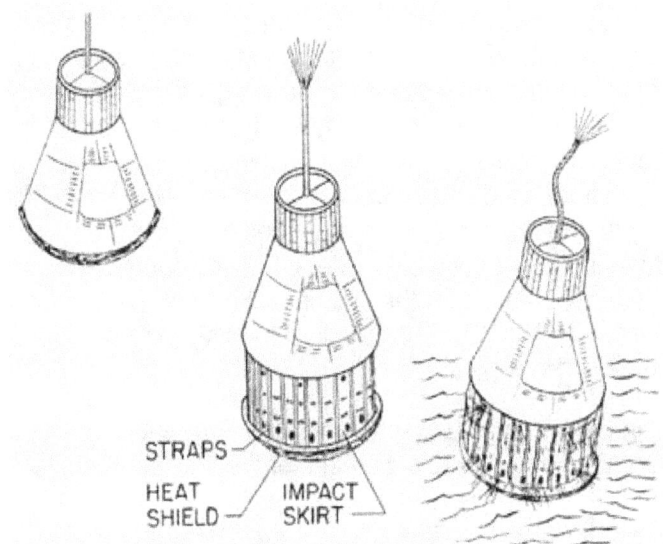

Figure 46. Impact attenuation.

February (during the month)

The orbital psychomotor tester qualification tests began.
> NASA Space Task Group, *Project Mercury [Quarterly] Status Report No. 10 for Period Ending April 30, 1961.*

Instruction was provided to the astronauts to develop techniques and procedures for using the personal parachute as an additional safety feature in the Mercury program. This parachute was only used during the Mercury-Redstone 3 (MR-3) mission manned by Alan Shepard.

> NASA Space Task Group, *Project Mercury [Quarterly] Status Report No. 9 for Period Ending January 31, 1961.*

March 2

Evaluation of the Mercury-Atlas 2 (MA-2) flight results disclosed that the spacecraft afterbody temperatures were somewhat lower than had been anticipated.
> NASA Space Task Group, *Project Mercury [Quarterly] Status Report No. 10 for Period Ending April 30, 1961.*

March 3

Factory roll-out inspection of Atlas launch vehicle No. 100-D was conducted at Convair-Astronautics. This launch vehicle was allocated for the Mercury-Atlas 3 (MA-3) mission.
> NASA Space Task Group, *Project Mercury [Quarterly] Status Report No. 10 for Period Ending April 30, 1961.*

March 6

"Detailed Test Objectives for NASA Mission MA-3" was published.
> Report, *Detailed Test Objectives for NASA Mission MA-3*, prepared by Mercury Space-Booster Program Office, The Aerospace Corporation, March 6, 1961.

March 6-7

The third in the series of development engineering inspections on Mercury spacecraft was held. At this time, spacecraft Nos. 12 and 15 were inspected, and some 50 requests for alterations were made.
> NASA Space Task Group, *Project Mercury [Quarterly] Status Report No. 10 for Period Ending April 30, 1961.*

March 7

Spacecraft No. 11 was delivered to Cape Canaveral for the Mercury-Redstone 4 (MR-4) ballistic manned (Grissom) flight.
> Data supplied by Ken Vogel, Mercury Project Office, MSC.

Redstone launch vehicle No. 5 was delivered to Cape Canaveral for the Mercury-Redstone, Booster Development flight (MR-BD).

> Data supplied by Ken Vogel, Mercury Project Office, MSC.

March 8

Spacecraft No. 10 was accepted and delivered to the McDonnell altitude test facility on March 31, 1961, for an orbital-flight environmental test.
> Data supplied by Ken Vogel, Mercury Project Office, MSC.

March 14

Atlas launch vehicle 100-D was delivered to Cape Canaveral for the Mercury-Atlas 3 (MA-3) mission. (See fig. 47.)
> Data supplied by Ken Vogel, Mercury Project Office, MSC.

Figure 47. Atlas launch vehicle 100-D delivered to Cape Canaveral for Mercury-Atlas 3 flight.

March 16

The Space Task Group recommended that the Department of Defense give consideration to assigning weather reconnaissance missions to the Air Weather Service preceding Mercury orbital missions beginning with Mercury-Atlas 4 (MA-4).

> Letter, Walter C. Williams, Associate Director of Project Mercury to Department of Defense Representative, Project Mercury Support Operations, subject: Weather Reconnaissance Flights in Support of Project Mercury, March 16, 1961.

Mercury spacecraft No. 10 was withdrawn from the flight program and was allocated to a ground test simulating orbital flight environmental conditions at the McDonnell plant site.

> Data supplied by Ken Vogel, Mercury Project Office, MSC.

The Space Task Group advised the Goddard Space Flight Center that for all Mercury orbital missions, beginning with Mercury-Atlas 3 (MA-3), trajectory data would be required for postflight analysis.

> Memo, Space Task Group to Goddard Space Flight Center, subject: Requirements for Project Mercury Postflight Computing at Goddard, Mar. 16, 1961.

Mission rules for Mercury-Atlas 3 (MA-3) were published. Revisions were issued on April 4, and April 20, 1961.

> Directive, subject: Mercury Control Center Countdown Flight Control and Overall Operations, MA-3, Apr. 20, 1961.

March 18

Little Joe 5A (LJ-5A), the sixth in the series of Little Joe missions, was launched from Wallops Island. This flight was intended to satisfy test objectives, which were not met previously because of the failure of the spacecraft to separate from the launch vehicle during the Little Joe 5 (LJ-5) mission flown on November 8, 1960. For reference, the purpose of this test was to demonstrate primarily the structural integrity of the spacecraft and the escape system during an escape maneuver initiated at the highest dynamic pressure anticipated during an Atlas launch for orbital flight. Little Joe 5A (LJ-5A) lifted off normally, but 19 seconds later the escape tower fired prematurely, a situation closely resembling the November 1960 flight. The signal to initiate the abort maneuver was given; and the launch vehicle-adapter clamp ring was released as intended, but the spacecraft remained on the launch vehicle since the escape motor was already expended. The separation was effected by using the retrorockets, but this command was transmitted before the flight had reached its apex, where separation had been planned. Therefore, the separation was rather violent. The parachutes did deploy at about 40,000 feet, and after recovery it was found that the spacecraft had actually incurred only superficial structural damage. In fact, this spacecraft was later used for the subsequent Little Joe 5B (LJ-5B) flight test. Test objectives of the Little Joe 5A (LJ-5A) were not met.
> Memo, Warren J. North to NASA Administrator, subject: Preliminary Flight Results, Little Joe 5A, March 20, 1961; NASA Space Task Group, *Project Mercury [Quarterly] Status Report No. 10 for Period Ending April 30, 1961*.

March 20

Between this date and April 13, 1961, Phase III of the spacecraft airdrop program was conducted. Primary objectives of the drops were to study further the spacecraft suitability and flotation capability after water impact. Six drops were made, but later (April 24-28, 1961) the tests were extended for two additional drops to monitor hard-surface landing effects. In the water phase of the program, spacecraft components under particular scrutiny were the lower pressure bulkhead and its capability to withstanding heat shield recontact without impairing flotation capability. Helicopters were used to make the drops.
> NASA Space Task Group, *Project Mercury [Quarterly] Status Report No. 10 for Period Ending April 30, 1961*.

Trajectory data for the Mercury-Redstone Booster-Development (MR-BD) flight test were forwarded by Marshall Space Flight Center to the Space Task Group and other interested organizations. The purpose of this flight test was to provide a final check of the launch vehicle system prior to the manned suborbital flights.

Letter, Marshall Space Flight Center to Space Task Group, *et al*, subject: Project
Mercury-Redstone: Trajectory Data for MR-BD, March 20, 1961.

March 21

The Mercury-Atlas Missile Range Projects Office, headed by Elmer H. Buller, was designated as a staff function of the Space Task Group Director's office.
Memo, Robert R. Gilruth for Space Task Group Staff, subject: Changes in Organization of the Space Task Group, March 21, 1961.

March 23

President John F. Kennedy advised Representative Overton Brooks (D-La.) that he had no intention "to subordinate" the space activities of the National Aeronautics and Space Administration to those of the military.
House of Representatives, Committee on Science and Astronautics, Press Release, April 2, 1961, with enclosures.

March 24

After analyzing launch vehicle behavior in the Mercury-Redstone 1A (MR-1A) and Mercury-Redstone 2 (MR-2), officials at the Marshall Space Flight Center and the Space Task Group were of the opinion that there were a number of problems that needed to be corrected prior to the advent of manned flight. The problems to be resolved included jet-vane vibration, instrumentation compartment vibration, failure of the thrust-controller system, and several other areas that needed attention. Many of these problems were studied by the personnel of engineering activities and proposed solutions were formulated. It was felt, however, that flight was necessary to verify the corrections and the Mercury-Redstone Booster Development test was scheduled and flown. All test objectives were met; as a result of this test, the launch vehicle was man-rated for the planned suborbital flights.
Memo, George Low to NASA Administrator, subject: Mercury Redstone Booster Development Test, March 27, 1961; NASA Space Task Group, *Project Mercury [Quarterly] Status Report No. 10 for Period Ending April 30, 1961.*

March 27

In a NASA Headquarters' note to editors of magazines and newspapers, a procedures and a deadline were established for submitting the applications of accredited correspondents to cover the Mercury-Redstone 3 (MR-3) flight mission. As of April 24, 1961, the deadline date, 350 correspondents were accredited to cover the launch, the first manned suborbital flight of Project Mercury.
NASA Note to Editors, Apr. 24, 1961.

March 30

Redstone launch vehicle No. 7 was delivered to Cape Canaveral for the Mercury-Redstone 3 (MR-3) mission.
>Data supplied by Ken Vogel, Mercury Project Office, MSC.

March 31

As of this date, all stations of NASA's world-wide Mercury tracking network were classed as being operational. An industrial team headed by the Western Electric Company turned over the $60,000,000 global network (figs. 48 and 49) to NASA in a formal ceremony later in the year.
>NASA Space Task Group, *Project Mercury [Quarterly] Status Report No. 10 for Period Ending April 30, 1961*; House Committee Print, subject: Aeronautical and Astronautical Events of 1961, June 7, 1962.

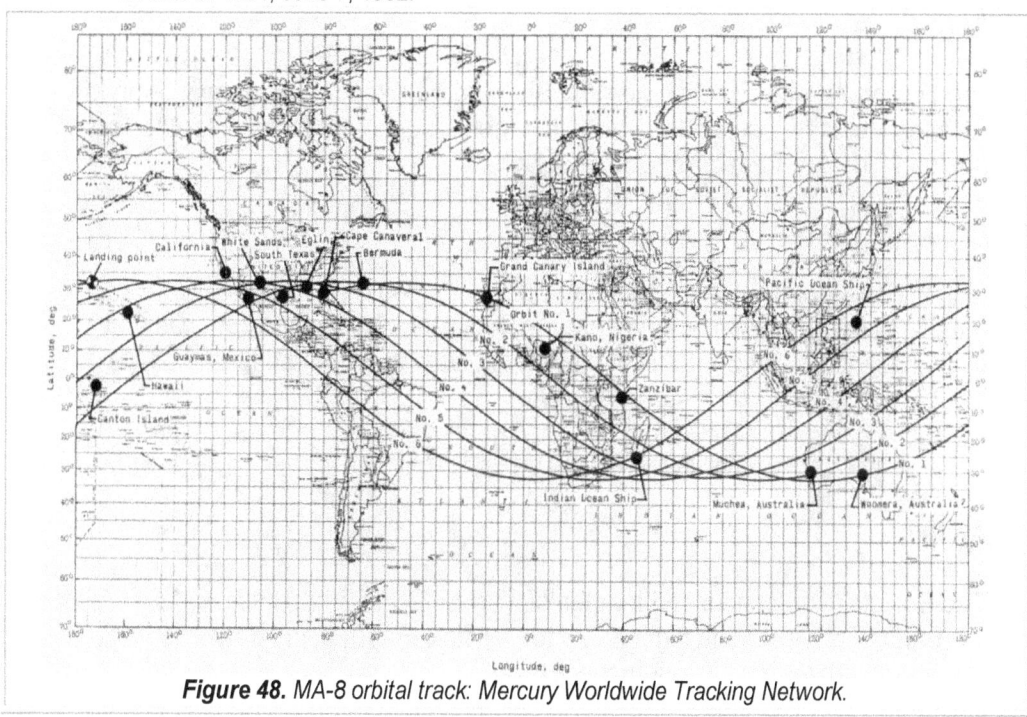

Figure 48. MA-8 orbital track: Mercury Worldwide Tracking Network.

Figure 49. Tracking site at Kano, Nigeria, Africa.

April 2

The first simulated orbital mission, with the spacecraft in the altitude chamber, was conducted.
>NASA Space Task Group, *Project Mercury [Quarterly] Status Report No. 10 for Period Ending April 30, 1961.*

April 3

To satisfy the national interest in Project Mercury, Robert R. Gilruth designated the Public Affairs Office as the point of contact for Space Task Group activities to supply information, within the limits of security, for news dissemination.
>Memo, Robert R. Gilruth to Space Task Group, subject: Public Affairs Activities, April 3, 1961.

April 4

John Glenn, Virgil Grissom, and Alan Shepard began a refresher course on the Aviation Medical Acceleration Laboratory centrifuge in preparation for the first manned Mercury-Redstone suborbital flight.
>House Committee Print, subject: Aeronautical and Astronautical Events of 1961, June 7, 1962.

Mercury spacecraft No. 14A was delivered to Wallops Island for the Little Joe 5B (LJ-5B) maximum dynamic-pressure abort mission. This spacecraft was first used in the Little Joe 5A (LJ-5A) mission and was then refitted for the LJ-5B flight.

>Data supplied by Ken Vogel, Mercury Project Office, MSC.

April 12

The Soviet Union announced that Major Yuri A. Gagarin had successfully orbited the Earth in a 108 minute flight in a 5 ton Vostok (East), the first man to make a successful orbital flight through space.
>House Committee, *Aeronautical and Astronautical Events of 1961*, June 7, 1962.

April 18

The United States Weather Bureau stated that funds in the amount of $200,000 would be required to support Project Mercury during the fiscal year of 1962.
>Letter, U.S. Department of Commerce, Weather Bureau, to Robert R. Gilruth, Director of Project Mercury, no subject: April 18, 1961.

April 20

Spacecraft, mission, and launch vehicle flight safety were reviewed by Space Task Group personnel in preparation for the Mercury-Redstone 3 (MR-3) mission.
> Mercury Control Center Countdown Flight Control and Overall Operations, MR-3, Apr. 20, 1961.

April 25

Mercury-Atlas 3 (MA-3) was launched from Cape Canaveral in an attempt to orbit the spacecraft with a "mechanical astronaut" aboard. After lift-off, the launch vehicle failed to roll to a 70 degree heading and to pitch over into the proper trajectory. The abort-sensing system activated the escape rockets prior to the launch vehicle's destruction by the range safety officer after approximately 40 seconds of flight that had attained an altitude of 16,400 feet. The spacecraft then coasted up to 24,000 feet, deployed its parachutes, and landed in the Atlantic Ocean 2,000 yards north of the launch pad. The spacecraft was recovered and was found to have incurred only superficial damage; it was then shipped to McDonnell for refitting.
> Report, subject: Mercury-Atlas No. 3 (MA-3), Memorandum Report for the Director, prepared by the Projects Engineering Branch, Space Task Group, April 28, 1961.

President Kennedy signed legislation making the Vice President of the United States the presiding officer of the National Aeronautics and Space Council.

> House Committee, *Aeronautical and Astronautical Events of 1961*, June 7, 1962.

April 28

Little Joe 5B (LJ-5B) was launched from Wallops Island to test the Mercury escape system under maximum dynamic pressure conditions. At the time of lift-off, one of the launch vehicle rocket motors did not ignite until after 4 seconds had elapsed. This delay caused the launch vehicle to pitch into a lower trajectory than had been planned, with a result that the abort maneuver experienced greater dynamic pressures than had been specified in the flight test plan. Other than this, all other sequential systems operated according to plan, and after landing, a normal helicopter recovery was accomplished. Thus, all test objectives were met and were actually exceeded because the spacecraft withstood the higher dynamic pressures.
> Memo, Warren J. North to NASA Administrator, subject: Little Joe 5B Launch, Apr. 28, 1961.

A simulated countdown for the first Mercury-Redstone manned suborbital flight (MR-3) was successfully completed.

> House Committee Print, subject: Aeronautical and Astronautical Events of 1961, June 7, 1962.

May 5

A document was issued regarding use of a Scout test vehicle (fig. 50) to evaluate the performance of the Mercury tracking and real-time computing system. NASA Headquarters tentatively approved the plan on May 24, 1961.
    Memo, Abe Siverstein to NASA Associate Administrator, subject: Use of Scout for Checkout of Mercury Network, May 24, 1961.

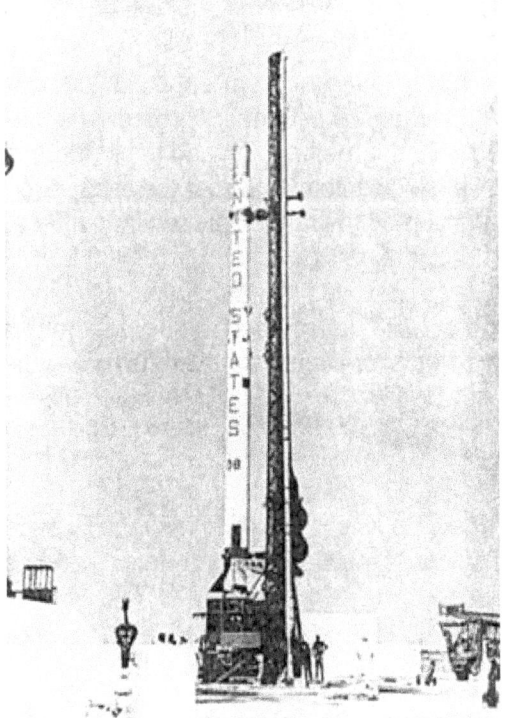

Figure 50. Scout launch vehicle proposed to test Mercury Worldwide Tracking Network.

1961 (during the year)

Prior to entering the operational phase of Project Mercury, a decision was made by Robert R. Gilruth and James E. Webb that the astronaut selected for each flight would have the right to name his spacecraft, which is in keeping with past traditions. Therefore, the astronaut advised Robert R. Gilruth of the name of the spacecraft which he had chosen (Freedom 7 in the case of the first flight) and Mr. Gilruth, in turn, advised Mr. Webb of the name. The Federal Communications Commission was also notified of the name since the spacecraft would be using communications frequencies controlled by the Commission.
    Information supplied by Lt. Col. John A. Power, Public Affairs Office, MSC, Aug. 5, 1963.

PART III (A)
Operational Phase of Project Mercury
May 5, 1961 through May 1962

1961

May 5

Mercury-Redstone 3 (MR-3), designated the Freedom 7, the first Mercury manned suborbital flight, was launched from Cape Canaveral, with astronaut Alan Shepard as the pilot. (See fig. 51.) The Redstone booster performed well during the boosted phase, although there were some vibrations, and cutoff was well within specified limits. After separation, Shepard exercised manual control of the spacecraft in the fly-by-wire and manual proportional modes. The attitude control system operated well, with few thruster fuel leaks. Reentry and landing were accomplished without any difficulty. During the flight, the spacecraft attained a maximum speed of 5,180 miles per hour, rose to an altitude of 116.5 statute miles, and landed 302 statute miles downrange from Cape Canaveral. (See fig. 52.) The pilot experienced a maximum of 6 g's during the booster acceleration phase and slightly less than 12 g's upon reentry. The duration of the flight was 15 minutes and 22 seconds, with weightlessness existing for approximately 5 minutes. Recovery operations were perfect, as helicopters were able visually to follow the descent of the spacecraft. Contact was made with the pilot two minutes after impact and recovery was initiated. (See fig. 53.) There was no damage to the spacecraft, and Shepard was in excellent condition. The first Mercury suborbital flight was a success.
> NASA Report, *Proceedings of a Conference on Results of the First U.S. Manned Suborbital Space Flight*, June 6, 1961.

May 8

Astronaut Alan Shepard, pilot of the Freedom 7 spacecraft (MR-3) was awarded NASA's Distinguished Service Medal by President John F. Kennedy in a ceremony at the White House.
> NASA, *The Space Flight of Astronaut Shepard and the Freedom Seven*.

May 11

Mercury spacecraft 8A was delivered to Cape Canaveral for the Mercury-Atlas 4 (MA-4) orbital unmanned (mechanical astronaut) mission.
> Data supplied by Ken Vogel, Mercury Project Office, MSC.

Figure 51. Mercury-Redstone 3: First manned suborbital space flight.

May 13

NASA submitted its legislative program for the 87th Congress (S. 1857 and H.R. 7115), asking for authority to lease property, authority to acquire patent releases, replacement of semiannual reports to Congress with an annual one, and authority to indemnify contractors against unusually hazardous risks.
House Committee, *Aeronautical and Astronautical Events of 1961*, June 7, 1962.

May 17

An Atlas investigation board was convened to study the cause of the Mercury-Atlas 3 (MA-3) mission launch vehicle failure. Several possible areas were considered, and three were isolated as probable causes based on a review of test data.

> Letter, B. A. Hohmann, Program Director, Mercury Space-Booster, Aerospace Corporation, to NASA Hq., subject: Transmittal of Mercury Atlas 100-D Investigation Board Status Report, June 14, 1961, with enclosures.

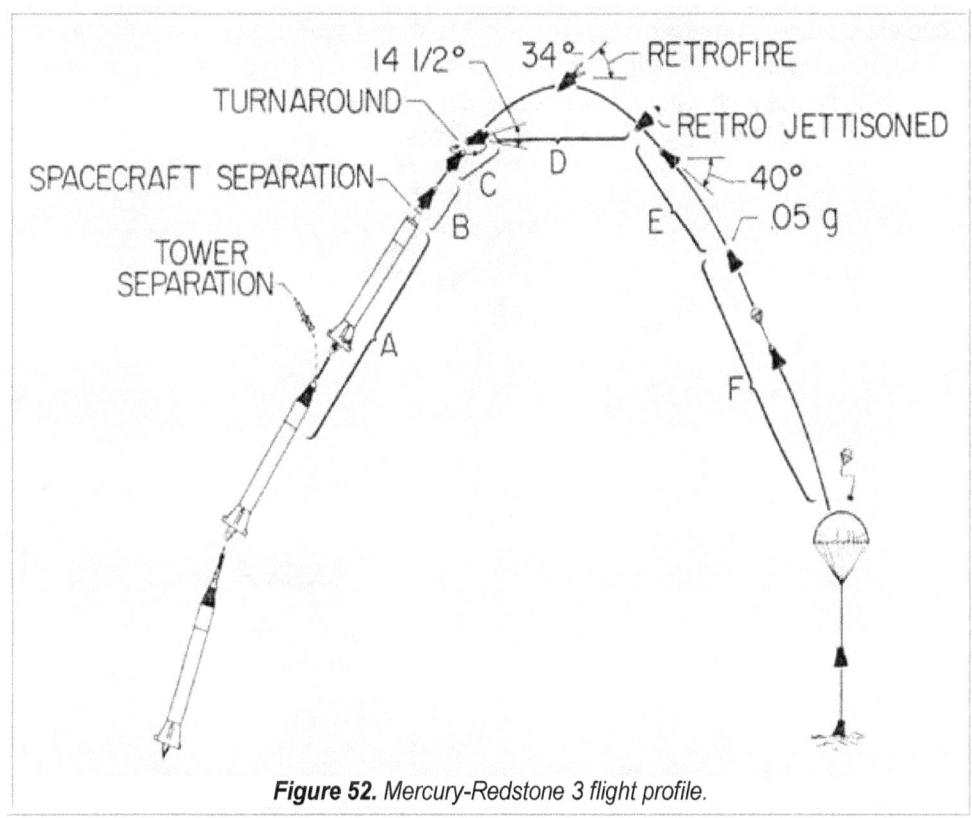

Figure 52. Mercury-Redstone 3 flight profile.

May 19

NASA Headquarters and the Space Task Group began a concerted effort in reviewing Mercury progress to identify technical developments that were potential inventions, discoveries, improvements, and innovations. This action was in keeping with the policy and concept of providing information on technical advances, within security limits and when appropriate, to other agencies of the government and to American industry.

> Memo, Glenn F. Bailey, Contracting Officer, Space Task Group, to J. M. Carson, Office of Patent Counsel, Langley, subject: Contract NAS 5-59 Inventions, Sept. 8, 1961.

May 23-24

The fourth development engineering inspection on Mercury spacecraft was held at McDonnell. Inspection activities were primarily centered on spacecraft No. 18, and some 45 requests for alterations were initiated.

Memo, Richard B. Ferguson to Assistant Chief for Mercury Support, Flight Systems Division, Space Task Group, subject: Trip to McDonnell to Attend 4th DEI, June 2, 1961.

May 25

President Kennedy, in a major message to Congress, called for a vastly accelerated space program based on a long-range national goal of landing a man on the moon and bringing him safely back to Earth. For this and associated projects in space technology, the President requested additional appropriations totaling $611 million for NASA and the Department of Defense.
House Committee, *Aeronautical and Astronautical Events of 1961*, June 7, 1962.

Figure 53. Freedom 7 returned by helicopter to USS Lake Champlain.

May 26-27

The first conference on the "Peaceful Uses of Space" was held at Tulsa, Oklahoma. A second conference on this subject was held at Seattle,

Washington, on May 8-10, 1962. In both instances, Robert R. Gilruth reported on the manned space flight aspect.
> *Proceedings of First National Conference on the Peaceful Uses of Space*, Tulsa, Oklahoma, May 26-27, 1961; *Proceedings of Second National Conference on the Peaceful Uses of Space*, Seattle, Washington, May 8-10, 1962.

May 26

Between this date and June 4, 1961, the Mercury spacecraft Freedom 7 (MR-3) was displayed at the Paris International Air Show. Some 650,000 visitors received the details on the spacecraft and on Shepard's suborbital flight.
> House Committee, *Aeronautical and Astronautical Events of 1961*, June 7, 1962.

May 29

Between this date and June 30, 1959, a centrifuge training program was conducted at the Aviation Medical Acceleration Laboratory directed entirely toward training the astronauts for the Mercury-Atlas orbital missions.
> NASA Space Task Group, *Project Mercury [Quarterly] Status Report No. 11 for Period Ending July 31, 1961*.

June 1

Prelaunch mission rules for Mercury-Atlas 4 (MA-4) were published.
> Letter, Walter C. Williams, Associate Director, Space Task Group, to Colonel Paul R. Wignall, Patrick AFB, Florida, subject: Transmittal of Prelaunch Mission Rules for MA-4, June 1, 1961.

June 6

Biomedical results of the Mercury-Redstone 3 (MR-3), Shepard's suborbital space flight, were reported in a Washington conference jointly sponsored by NASA, National Institute of Health, and the National Academy of Sciences.
> Memo, George Low, NASA Hq., to William H. Allen, subject: Conference on Medical Results of the First U.S. Manned Suborbital Flight, June 16, 1961.

June 8

Mercury-Atlas 4 (MA-4) recovery requirements were published.
> Letter, Walter C. Williams, Space Task Group, to Commander, DESFLOTFOUR, no subject, June 8, 1961, with enclosures.

June 12

Redstone launch vehicle No. 8 was delivered to Cape Canaveral for the Mercury-Redstone 4 (MR-4) suborbital flight mission.
> Data supplied by Ken Vogel, Mercury Project Office, MSC.

June 13

The Space Task Group forwarded to NASA Headquarters the details for the Mercury-Scout instrumentation system. This mission was to check the operational effectiveness of the Mercury global tracking network.
    Memo, Warren J. North to NASA Deputy Director, Space Flight Programs, subject: Mercury Status Items for Project Review Meeting, June 27, 1961, June 22, 1961.

June 13-25

The Freedom 7 (MR-3) spacecraft was viewed by approximately 750,000 visitors at the Rassegna International Electronic and Nuclear Fair at Rome, Italy.
    House Committee, *Aeronautical and Astronautical Events of 1961*, June 7, 1962.

June 16

An Ad Hoc Task Group reported to NASA the results of its studies to determine the main problems, the pacing items, and the major decisions required to accomplish the manned lunar landing mission. The direct ascent method was studied intensively with much less attention given to the rendezvous method.
    House Committee, *Aeronautical and Astronautical Events of 1961*, June 7, 1962.

June 21

Between the cited date and July 15, 1961, as a part of the Mercury-Atlas animal program, chimpanzees received training in acclimation to noise and vibration and to centrifuge runs at the University of Southern California. Two of the animals flew parabolas in a C-131 aircraft for weightlessness training. The animals were also trained in advance psychomotor problems.
    NASA Space Task Group, *Project Mercury [Quarterly] Status Report No. 11 for Period Ending July 31, 1961*.

June 22

Mercury-Redstone 4 (MR-4) recovery requirements were forwarded by the Space Task Group to the Navy.
    Letter, Walter C. Williams, Space Task Group, to Commander, DESFLOTFOUR, subject: Mercury-Redstone No. 4 Recovery Requirements, June 22, 1961.

The Redstone booster for the Mercury-Redstone 4 (MR-4) manned suborbital flight mission was erected on Pad 5, at Cape Canaveral.

    House Committee, *Aeronautical and Astronautical Events of 1961*, June 7, 1962.

June 24

Modifications were made to the spacecraft designated for the second manned suborbital Mercury flight. An observation window replaced two view ports and an improved manual control system was installed.
House Committee, *Aeronautical and Astronautical Events of 1961*, June 7, 1962.

June 28

Using spacecraft No. 5, a spacecraft seaworthiness test was conducted 65 miles east of Wallops Island. Sea conditions varied with 2 to 4 foot ground swells and wave heights of from 1 to 2 feet. Spacecraft flotation characteristics were found to be quite satisfactory.
NASA Space Task Group, *Project Mercury [Quarterly] Status Report No. 11 for Period Ending July 31, 1961*.

Tracking network requirements for the Mercury extended range or 1 day mission were discussed between Space Task Group and Goddard Space Flight Center personnel.

NASA Space Task Group, *Project Mercury [Quarterly] Status Report No. 11 for Period Ending July 31, 1961*.

June 29-30

Factory roll-out inspection of Atlas launch vehicle 88-D, designated for the Mercury-Atlas 4 (MA-4) mission, was conducted at Convair.
NASA Space Task Group, *Project Mercury [Quarterly] Status Report No. 11 for Period Ending July 31, 1961*.

Personnel strength of the Space Task Group was 794.

Statistics supplied by Kathryn Walker, Personnel Division, MSC.

July 1

Responsibility for the operation of the Mercury global network was assigned to the Goddard Space Flight Center. During active mission periods, network control would revert to Space Task Group personnel.
NASA Space Task Group, *Project Mercury [Quarterly] Status Report No. 11 for Period Ending July 31, 1961*.

July 11

Key personnel assignments were made by Walter C. Williams, Project Mercury Operations Officer, for the Mercury-Redstone 4 (MR-4) manned suborbital flight mission. These appointments included on-site liaison and consultation, public affairs, photo couriers, and technical observers. Stations covered were Mercury

Control Center (fig. 54), Atlantic Missile Range Central Control, landing area aircraft carrier, supporting destroyers, support aircraft, and Base Operations at Patrick Air Force Base.

> Letter, Walter C. Williams to Department of Navy, subject: MR-4 Recovery Requirements, July 11, 1961.

Figure 54. Key personnel in Mercury control center at Cape Canaveral: L to R, Walter C. Williams, Flight Director; John A. (Shorty) Powers, Mission Narrator; Christopher C. Kraft, Flight Director.

July 13

Mercury-Redstone 4 (MR-4) manned suborbital flight mission rules were published.

> Memo, E. F. Kranz, Flight Control Branch, Space Task Group, subject: MR-4 Mission Rules, July 13, 1961.

The Redstone launch vehicle designated for the Mercury-Redstone 6 (MR-6) mission was static tested at the Marshall Space Flight Center to ensure satisfactory operation of the turbopump assembly.

> House Committee, *Aeronautical and Astronautical Events of 1961*, June 7, 1962.

July 13-15

A spacecraft, launch vehicle, and mission flight safety review was held in preparation for the Mercury-Redstone 4 (MR-4) manned suborbital flight mission.

> NASA Space Task Group, *Project Mercury [Quarterly] Status Report No. 11 for Period Ending July 31, 1961.*

July 15

Atlas launch vehicle 88-D was delivered to Cape Canaveral for the Mercury-Atlas 4 (MA-4) mission.
    Data supplied by Ken Vogel, Mercury Project Office, MSC.

July 18-19

Two attempts were made to launch Mercury-Redstone 4 (MR-4) with astronaut Virgil Grissom aboard the spacecraft, but unfavorable weather forced mission postponement.
    House Committee, *Aeronautical and Astronautical Events of 1961*, June 7, 1962.

July 21

Mercury-Redstone 4 (MR-4), designated Liberty Bell 7, the second Mercury manned suborbital flight, was launched from Cape Canaveral with astronaut Virgil Grissom as the pilot. From lift-off to reentry, operational sequences were similar to those of the first manned suborbital flight. In the ballistic trajectory, the spacecraft reached a peak altitude of 118 statute miles and landed 303 statute miles downrange from Cape Canaveral. Grissom's flight experience was similar to Shepard's in that there was a 5 minute period of weightlessness, and neither reported any ill effects resulting from this condition. The MR-4 pilot also found it easy to control his spacecraft attitude in the manual mode of operation. The spacecraft was lost during the recovery operations, when the explosive side egress hatch activated prematurely while Grissom was awaiting helicopter pickup. The astronaut egressed immediately and was retrieved after swimming in the water 3 or 4 minutes. With this second successful suborbital flight, the Space Task Group felt there was nothing further to be gained from this phase of Project Mercury, and the remaining Redstone launch vehicle flights were canceled.
    NASA Space Task Group Report, *Results of the Second U.S. Manned Suborbital Space Flight*, July 21, 1961.

July 22

Astronaut Virgil Grissom, pilot of the MR-4 Liberty Bell 7, was awarded the NASA Distinguished Service Medal by NASA Administrator James Webb at the conclusion of the MR-4 press conference held at Cape Canaveral.
    NASA, *Liberty Bell 7*, July 21, 1961.

July 27-28

After the 2-man space concept (later designated Project Gemini) was introduced in May 1961, a briefing between McDonnell and NASA personnel was held on the matter. As a result of this meeting, space flight design effort was concentrated on the 18-orbit 1-man Mercury and on a 2-man spacecraft capable of advanced missions.

Notes on the early history of Project Gemini, prepared by McDonnell Aircraft Corporation, undated.

July 31

Between the cited date and September 15, 1961, the astronaut centrifuge training program at the Aviation Medical Acceleration Laboratory was directed entirely toward the Mercury-Atlas orbital missions.
NASA Space Task Group, *Project Mercury [Quarterly] Status Report No. 11 for Period Ending July 31, 1961.*

August 1-3

Seaworthiness characteristics of the operational Mercury spacecraft were evaluated. Conditions during the test varied from ground swells of 5 to 15 feet, wave heights of 2 to 10 feet, and winds of 6 to 20 knots. The test lasted for 33 hours and was quite successful.
NASA Space Task Group, *Project Mercury [Quarterly] Status Report No. 12 for Period Ending October 31, 1961.*

August 6

U.S.S.R. launched Vostok II into orbit, carrying Major Gherman S. Titov. The spacecraft weighed 13 pounds more than Vostok I (April 12), and the progress of Cosmonaut Titov's flight was reported continuously on Radio Moscow.
House Committee, *Aeronautical and Astronautical Events of 1961*, June 7, 1962.

August 9

Retrofire-from-orbit mission rules were published for the unmanned Mercury-Atlas 4 (MA-4) orbital flight.
Memo, Flight Operations Division, Space Task Group, to Associate Director, subject: Retrofire from Orbit Mission Rules for MA-4, Aug. 9, 1961.

Key personnel operational assignments for the Mercury-Atlas 4 (MA-4) unmanned orbital mission were made by the Space Task Group.

Letter, Walter C. Williams, Space Task Group, to Commander, DESFLOTFOUR, subject: NASA Personnel Assignments for MA-4 Test, Aug. 9, 1961, with inclosures.

August 13

Spacecraft No. 15 was delivered to Cape Canaveral, but was returned to McDonnell to be reconfigured to the orbital-manned 1-day mission and tentatively assigned for Mercury-Atlas 10 (MA-10). Redesign was completed, and the spacecraft, then designated number 15A (later redesignated 15B), was delivered to Cape Canaveral on November 16, 1962.
Data supplied by Ken Vogel, Mercury Project Office, MSC.

August 18

NASA Headquarters publicly announced that an analysis of Project Mercury suborbital data indicated that all objectives of that phase of the program had been achieved and no further Mercury-Redstone flights were planned.
    House Committee, *Aeronautical and Astronautical Events of 1961*, June 7, 1962.

August 22

Between the cited date and September 12, 1961, mission, spacecraft, and launch vehicle flight safety reviews were held for the unmanned Mercury-Atlas 4 (MA-4) orbital flight.
    NASA Space Task Group, *Project Mercury [Quarterly] Status Report No. 12 for Period Ending October 31, 1961*.

August 24

Mercury-Atlas 4 (MA-4) unmanned orbital flight was postponed.
    House Committee, *Aeronautical and Astronautical Events of 1961*, June 7, 1962.

August 25

Explorer XIII, designed in part to measure the effects of micrometeoroids on spaceflight, failed to meet expectations, thereby necessitating further tests in this area.
    Goddard Space Flight Center Chart, Satellites and Space Probe Projects as of July, 1962.

August 27

Spacecraft No. 13 was shipped to Cape Canaveral. This particular vehicle was designated for the first manned Mercury-Atlas orbital flight (MA-6, Glenn). Test and checkout work on the spacecraft was started immediately.
    Data supplied by Ken Vogel, Mercury Project Office, MSC.

August 30

An investigation was conducted as a result of the premature activation of the Mercury-Redstone 4 (MR-4) explosive egress hatch. Tests were initiated in an environment more severe than had been conducted in prelaunch activities and tests, but no premature firings occurred. As a backup, McDonnell was asked to design a mechanical-type hatch. The model weighed some 60 pounds more than the explosive type, so other methods had to be sought to prevent any recurrence of the incident. A procedure was initiated which stipulated that the firing plunger safety pin would be left in place until the helicopter hook was attached to the spacecraft and tension was applied to the recovery cable.
    Memo, Warren J. North to NASA Associate Administrator, subject: Report of Investigations, Aug. 30, 1961.

August (during the month)

A NASA site selection team, headed by John F. Parsons, Associate Director of the Ames Research Center, toured possible sites for the permanent location of a manned spacecraft center. The team graded the capabilities of these locations in meeting 10 specified requirements of the new center. These were: (1) available facilities for advanced scientific study; (2) power facilities and utilities; (3) water supply; (4) temperature climate; (5) adequate housing for center personnel; (6) at least 1,000 acres of land for the installation; (7) industrial facilities available; (8) transportation facilities, including water transportation for shipping cumbersome space vehicles by barge; (9) a first-class, all-weather jet service airport; and (10) local cultural and recreational assets. Sites considered were: Tampa, Florida; Jacksonville, Florida; New Orleans, Louisiana; Baton Rouge, Louisiana; Shreveport, Louisiana; Houston, Texas; Beaumont, Texas; Corpus Christi, Texas; Victoria, Texas; St. Louis, Missouri; Los Angeles, California; Berkeley, California; San Diego, California; Richmond, California; Moffett Field, California; San Francisco, California; Bogalusa, Louisiana; Liberty, Texas; Harlingen, Texas; and Boston, Massachusetts.
>House Committee, *Aeronautical and Astronautical Events of 1961*, June 7, 1962; Newport News, Virginia, *Daily Press*, Sept. 13, 1961; Information supplied by I. Edward Campagna, Facilities Division, MSC, June 16, 1963.

August 5 to October 12

A series of environmental tests was conducted on the spacecraft explosive egress hatch because of the difficulties experienced during the Mercury-Redstone 4 (MR-4) mission.
>NASA Space Task Group, *Project Mercury [Quarterly] Status Report No. 12 for Period Ending October 31, 1961*.

September 5, 9 and 14

Three rocket sled tests were conducted at the Naval Ordnance Test Station, China Lake, California, to study the detailed launch vehicle-spacecraft, clamp-ring separation. From run to run, minor modifications were made, and by the third run the separation action was perfected.
>NASA Space Task Group, *Project Mercury [Quarterly] Status Report No. 12 for Period Ending October 31, 1961*.

September 8

A report was made on possible technical advances as a result of the Mercury development program. A few of these are listed: (1) attenuation of impact force from astronaut couch by using crushable honeycomb structure; (2) interchangeable couch configuration for Mercury spacecraft; (3) modified tower clamp ring to improve stability in abort attitude; (4) hydrogen peroxide thrust chamber improvements; (5) oxygen pressure transducer improvements; (6) de-

stabilization flap to prevent spacecraft wrong attitude reentry; (7) Mercury spacecraft landing bag design; and (8) multi-nozzle rockets.
> Memo, Glenn F. Bailey, Contracting Officer, Space Task Group, to J. M. Carson, Office of Patent Counsel, Langley, subject: Contract NAS 5-59 Inventions, Sept. 8, 1961.

September 13

Mercury-Atlas 4 (MA-4) was launched from Cape Canaveral with special vibration and noise instrumentation and a mechanical crewman simulator aboard in addition to the normal spacecraft equipment. This was the first Mercury spacecraft to attain an earth orbit. The orbital apogee was 123 nautical miles and the perigee was 86 nautical miles. After one orbit, the spacecraft's orbital timing device triggered the retrograde rockets, and the spacecraft splashed in the Atlantic Ocean 161 miles east of Bermuda. Recovery was made by the USS Decatur. During the flight, only three slight deviations were noted - a small leak in the oxygen system; loss of voice contact over Australia; and the failure of an inverter in the environmental control system. Overall, the flight was highly successful: the Atlas booster performed well and demonstrated that it was ready for the manned flight, the spacecraft systems operated well, and the Mercury global tracking network and telemetry operated in an excellent manner and was ready to support manned orbital flight. (See fig. 55.)
> Memo, George Low to NASA Administrator, subject: Preliminary Results of MA-4 Flight, Sept. 15, 1961.

September 18

Mission rules for the Mercury-Atlas 5 (MA-5) orbital flight were published. Revisions were issued on October 16 and 25, 1961, and November 11, 1961.
> Memo, Eugene F. Kranz, Flight Control Operations, to J. A. Chamberlin, Space Task Group, subject: Mission Rules, MA-5/9, Nov. 11, 1961.

September 19

James Webb, NASA Administrator, announced that the new NASA center for manned space flight would be constructed upon a 1,000 acre site donated by Rice University, southeast of Houston, in Harris County, Texas. The Space Task Group would move from Langley Field to Houston, Texas.
> Memo, Robert R. Gilruth to Space Task Group Staff, subject: Location of New Site for Space Task Group, Sept. 19, 1961, with enclosures.

September 20

Robert R. Gilruth and other officials of the Space Task Group surveyed the Houston, Texas, area to seek temporary operational quarters while the permanent installation was being constructed.
> House Committee, *Aeronautical and Astronautical Events of 1961*, June 7, 1962

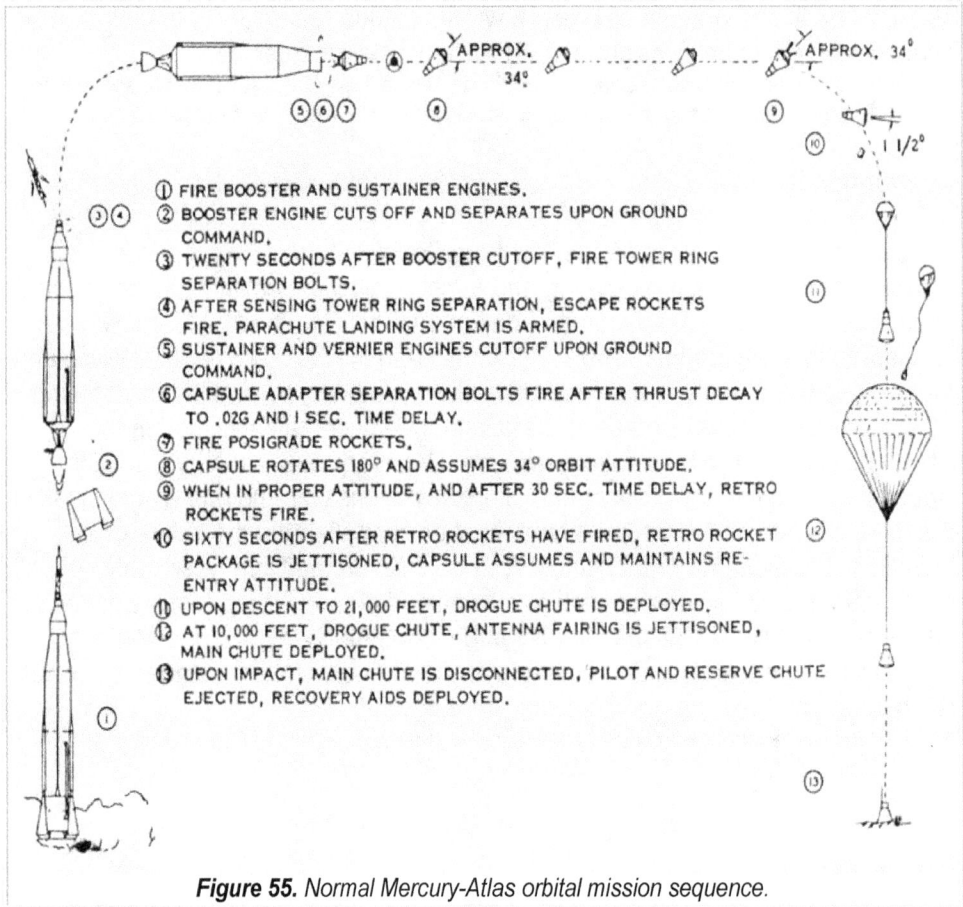

Figure 55. Normal Mercury-Atlas orbital mission sequence.

September 21

D. Brainerd Holmes was appointed NASA's Director of Manned Space Flight Programs. As general manager of Radio Corporation of America's Major Defense Systems Division, Holmes had been project manager of the Ballistic Missile Early Warning System. Congressman G. P. Miller (D.-Calif.) succeeded the recently deceased Congressman Overton Brooks of Louisiana as chairman of the House Committee on Science and Astronautics.
  House Committee, *Aeronautical and Astronautical Events of 1961*, June 7, 1962.

September 22

The Space Task Group announced that a 30-inch diameter balloon would be installed in the Mercury spacecraft to allow for ship recovery should the helicopter be forced to drop the spacecraft, as happened during the Mercury-Redstone 4 (MR-4) recovery operations.
  House Committee, *Aeronautical and Astronautical Events of 1961*, June 7, 1962.

September 24

NASA Administrator Webb announced major organizational changes and top-level appointments to become effective November 1. The reorganization should provide a clearer focus on major programs and allow center directors to have a louder voice in policy making. The new appointments included the following Directors of major program offices: Ira H. Abbott, Office of Advanced Research and Technology; Homer E. Newell, Office of Space Sciences; D. Brainerd Holmes, Office of Manned Space Flight; and an as yet unnamed Director of Office of Applications Programs. Also, Thomas F. Dixon was appointed Deputy Associate Administrator; Abe Silverstein was named Director of the Lewis Research Center, and Robert R. Gilruth was chosen Director of the Manned Spacecraft Center.

    House Committee, *Aeronautical and Astronautical Events of 1961*, June 7, 1962.

Evaluation of the inflatable flotation collar, attached by ground personnel to sustain spacecraft buoyancy during recovery operations, was completed. (See fig. 56.)

    NASA Space Task Group, *Project Mercury [Quarterly] Status Report No. 12 for Period Ending October 31, 1961*.

Figure 56. Auxiliary flotation collar.

October 1

Factory roll-out inspection of Atlas booster No. 93-D was conducted at Convair. This booster was designated for the Mercury-Atlas 5 (MA-5) mission.
    NASA Space Task Group, *Project Mercury [Quarterly] Status Report No. 12 for Period Ending October 31, 1961*.

October 9

Atlas booster No. 93-D was delivered to Cape Canaveral for the Mercury-Atlas 5 (MA-5) orbital flight mission.
>Data supplied by Ken Vogel, Mercury Project Office, MSC.

October 13

NASA Headquarters approved construction projects for a permanent manned spacecraft center installation at Clear Lake, southeast of Houston, Texas. Buildings to be constructed included an auditorium, project management, cafeteria, flight operations and life systems, life systems laboratory, technical services, technical services shop, central data processing, structures laboratory, research and development offices and laboratory, equipment evaluation laboratory, support offices, support warehouses and offices, and project test laboratory.
>Information extracted from Monthly Progress Report, Manned Spacecraft Center Facilities Division, August 1962; Notes, subject: Manned Spacecraft Center Building Facility Requirements, Oct. 13, 1961.

October 20

The Mercury-Atlas 5 (MA-5) data aquisition plan was published by the Mercury Data Coordination Office of the Space Task Group's Flight Operations Division.
>Plan, subject: MA-5 Data Acquisition Plan, prepared by Mercury Data Coordination Office, Flight Operations Division, Space Task Group, Oct. 20, 1961.

October 23

Freedom 7, the Mercury-Redstone 3 (MR-3) spacecraft, was presented by NASA to the National Air Museum of the Smithsonian Institution.
>House Committee, *Aeronautical and Astronautical Events of 1961*, June 7, 1962.

October 25

NASA Headquarters officially approved the Mercury extended range or 1-day mission program.
>NASA-MSC Report, *Project Mercury [Quarterly] Status Report No. 13 for Period Ending January 31, 1962.*

October 26-27

Ship retrieval tests were conducted to establish procedures for recovery of a manned Mercury spacecraft. No difficulties were encountered.
>NASA Space Task Group, *Project Mercury [Quarterly] Status Report No. 12 for Period Ending October 31, 1961.*

October 29

An announcement was made that a Mercury-Scout launch would be made to verify the readiness of the world-wide Mercury Tracking network to handle further orbital flights.
House Committee, *Aeronautical and Astronautical Events of 1961*, June 7, 1962.

October (during the month)

Spacecraft 12 was delivered to Cape Canaveral as a backup for the MA-8 mission (six-orbit flight), but immediate consideration was given for its modification to the Mercury extended range or 1-day mission. The capsule was returned to McDonnell, reconfigured and stored.
Data supplied by Ken Vogel, Mercury Project Office, MSC.

November 1

An attempt was made to launch Mercury-Scout 1 (MS-1) into orbit with a communications package further to qualify the radar tracking of the Mercury global network prior to manned orbital flight. Shortly after lift-off, the launch vehicle developed erratic motions and attending high aerodynamic loads, and was destroyed by the Range Safety Officer after 43 seconds of flight. No further attempts were planned. The Mercury-Atlas 4 (MA-4) mission and the successful Mercury-Atlas 5 (MA-5), flown on November 29, 1961, disclosed that the network met all requirements.
Memo, George Low to NASA Director, Office of Manned Space Flight, subject: Dynamic Checkout of the Mercury Ground Network with Mercury-Scout, Nov. 8, 1961.

The Space Task Group, the organization charged with directing Project Mercury and other manned spaceflight programs, was redesignated the Manned Spacecraft Center, with Robert R. Gilruth as Director.

Memo, Paul E. Purser to MSC Employees, subject: Designation of Space Task Group as "Manned Spacecraft Center," Nov. 1, 1961.

November 15

Mercury spacecraft No. 18 was delivered to Cape Canaveral for the second manned (Carpenter) orbital flight, Mercury-Atlas 7 (MA-7).
Data supplied by Ken Vogel, Mercury Project Office, MSC.

November 19

Factory roll-out inspection of Atlas launch vehicle 109-D was conducted. (See fig. 57.) This booster was designated for the Mercury-Atlas 6 (MA-6) mission, the first manned orbital space flight.
NASA-MSC Report, *Project Mercury [Quarterly] Status Report No. 13 for Period Ending January 31, 1962.*

Figure 57. Production of Atlas launch vehicles at Convair Astronautics plant at Sorrento, Calif.

November 29

For the Mercury-Atlas 5 (MA-5) orbital mission, the Mercury astronauts were assigned as spacecraft communicators at six of the Mercury global network tracking stations.
> NASA-MSC Report, *Project Mercury [Quarterly] Status Report No. 13 for Period Ending January 31, 1962.*

Mercury-Atlas 5 (MA-5), the second and final orbital qualification of the spacecraft prior to manned flight, was launched from Cape Canaveral with Enos, a 37.5 pound chimpanzee, aboard. (See fig. 58.) Scheduled for three orbits, the spacecraft was returned to earth after two orbits due to the failure of a roll reaction jet and to the overheating of an inverter in the electrical system. Both of these difficulties could have been corrected had an astronaut been aboard. The spacecraft was recovered 255 miles southeast of Bermuda by the *USS Stormes*. During the flight, the chimpanzee performed psychomotor duties and upon recovery was found to be in excellent physical condition. The flight was termed highly successful and the Mercury spacecraft well qualified to support manned orbital flight.

> NASA-MSC Report, *Project Mercury [Quarterly] Status Report No. 13 for Period Ending January 31, 1962.*

Astronaut John Glenn was selected as the pilot for the first manned orbital flight, with Scott Carpenter as backup pilot. Immediately, training was started to ready these two astronauts for the mission. The five remaining astronauts concentrated their efforts on various engineering and operational groups of the Manned Spacecraft Center in preparation for the mission.

NASA-MSC Report, *Project Mercury [Quarterly] Status Report No. 13 for Period Ending January 31, 1962.*

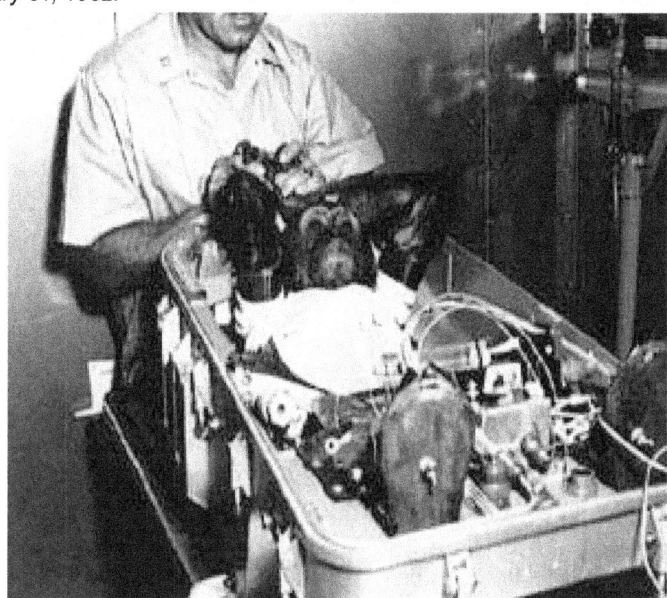

Figure 58. *Chimpanzee, "Enos," flown in Mercury-Atlas 5 two-orbit mission.*

November 30

Atlas launch vehicle 109-D was delivered to Cape Canaveral for the Mercury-Atlas 6 (MA-6) first manned orbital mission.
Data supplied by Ken Vogel, Mercury Project Office, MSC.

December 6

NASA Headquarters announced that the first Mercury manned orbital flight was scheduled for early 1962. This decision was made when the Mercury-Atlas 5 (MA-5) mission data indicated that the spacecraft system, launch vehicle, and tracking network were ready.
House Committee, *Aeronautical and Astronautical Events of 1961,* June 7, 1962.

In a joint ceremony, astronauts Alan Shepard and Virgil Grissom were awarded the first Astronaut Wings by their respective services.

House Committee, *Aeronautical and Astronautical Events of 1961,* June 7, 1962.

December 7

Plans for the development of a 2-man Mercury spacecraft were announced by Robert R. Gilruth, Director of the Manned Spacecraft Center. On January 3, 1962, this program was designated Project Gemini.
> House Committee, *Aeronautical and Astronautical Events of 1961*, June 7, 1962; NASA Historical Office, *Aerospace Chronology*, Jan. 1962.

December 11

A contract was awarded by the Army Corps of Engineers to a team headed by Brown and Root, Incorporated, for design of a major portion of the permanent facilities to be constructed for the Manned Spacecraft Center.
> House Committee, *Aeronautical and Astronautical Events of 1961*, June 7, 1962.

December 11-13

Spacecraft egress exercises were conducted for the astronauts in the Back River near Langley Field. This training was especially conducted for the pilots selected for the manned orbital mission and for helicopter recovery teams. The astronauts made both top and side hatch egresses from the spacecraft and no problems were encountered.
> NASA-MSC Report, *Project Mercury [Quarterly] Status Report No. 13 for Period Ending January 31, 1962.*

December 12

Spacecraft ultimate pressure tests to 20 pounds per square inch were conducted, and subsequent inspection disclosed there was no structural damage, deformation or failure.
> NASA-MSC Report, *Project Mercury [Quarterly] Status Report No. 13 for Period Ending January 31, 1962.*

December 14

Walter C. Williams told a University of Houston audience at Houston, Texas, that the Mercury spacecraft had served and would continue to serve as a test bed for developing orbital flight techniques and hardware for more ambitious space programs.
> Speech by Walter C. Williams, Deputy Director, MSC, Dec. 14, 1961.

December 14-18

Two Mercury spacecraft solid bottom (no impact bag) water drop tests were made. Subsequent inspections of the spacecraft structure and ablation heat shield disclosed no structural damage.
> NASA-MSC Report, *Project Mercury [Quarterly] Status Report No. 13 for Period Ending January 31, 1962.*

December 18

Spacecraft external pressure tests were conducted at pressures up to 15 pounds per square inch. Bulkhead deflection was slight and well within tolerable limits.
>NASA-MSC Report, *Project Mercury [Quarterly] Status Report No. 13 for Period Ending January 31, 1962.*

December 29

The appointments of Dr. Joseph F. Shea as Deputy Director for Systems Engineering, Office of Manned Space Flight at NASA Headquarters, and Dr. Arthur Rudolph as Assistant Director of Systems Engineering was announced. Dr. Rudolph would serve as liaison between vehicle development at Marshall Space Flight Center and the Manned Spacecraft Center in Houston.
>House Committee, *Aeronautical and Astronautical Events of 1961*, June 7, 1962.

December 31

Personnel strength of the Manned Spacecraft Center was 1,152.
>Statistics supplied by Katheryn Walker, Personnel Office, MSC.

Senior officials from NASA Headquarters, Marshall Space Flight Center, and the Manned Spacecraft Center will sit on a Management Council to insure the orderly and timely progress in the manned space flight programs. The Council under the chairmanship of D. Brainerd Holmes will meet at least once a month to identify and resolve problems as early as possible and to coordinate interface problems between the various Offices.

>House Committee, *Aeronautical and Astronautical Events of 1961*, June 7, 1962.

1962

January 1

A survey was performed at the Manned Spacecraft Center to ascertain the number of personnel who intended to move with the Center from Langley Field to Houston, Texas. Only 84 personnel indicated they would not make the move.
>Memo, Director of Personnel, MSC, to Philip H. Whitbeck, subject: Status Report for the Personnel Office, Jan. 26, 1962.

January 3

Exercises were held at the Lynnhaven Roads Anchorage near Norfolk, Virginia, to determine the feasibility of using the auxiliary flotation collar in recovery operations. The tests were successful and the collar was adopted.

NASA-MSC Report, *Project Mercury [Quarterly] Status Report No. 13 for Period Ending January 31, 1962.*

Flight controllers, excluding the medical monitors, were given a final briefing prior to deployment to remote sites for the Mercury-Atlas 6 (MA-6) mission.

NASA-MSC Report, *Project Mercury [Quarterly] Status Report No. 13 for Period Ending January 31, 1962.*

January 15

Organization and staffing of the Manned Spacecraft Center's Mercury Project Office was completed. Major organizational division of this staff element included Office of Project Manager, Project Engineering Office, Project Engineering Field Office (duty station at Cape Canaveral), Engineering Operations Office, and Engineering Data and Measurement Office. Kenneth Kleinknecht was appointed Manager of Project Mercury.

MSC Announcement No. 92-2 by Robert R. Gilruth, Director, subject: Establishment of Mercury Project Office, Jan. 15, 1962.

January 15-17

Recovery area swimmers were trained at the Pensacola Naval Air Station, Florida, for use in the Mercury-Atlas 6 (MA-6) manned orbital mission. (See fig. 59.) Instruction included films, briefings, auxiliary flotation collar deployment, and jumps from a helicopter.

NASA-MSC Report, *Project Mercury [Quarterly] Status Report No. 13 for Period Ending January 31, 1962.*

January 16

Spacecraft 16 was delivered to Cape Canaveral for the third manned (Schirra) orbital flight, Mercury-Atlas 8 (MA-8).

Data supplied by Ken Vogel, Mercury Project Office, MSC.

January 23

Robert R. Gilruth, Director of the Manned Spacecraft Center, was awarded the Louis W. Hill Space Transportation Award by the Institute of Aerospace Sciences for his "outstanding leadership in technical development of spacecraft for manned space flight."

NASA Historical Office, *Aerospace Chronology,* Jan. 1962.

Figure 59. Scuba divers prepare for recovery of Mercury spacecraft.

January 27

The Mercury-Atlas 6 (MA-6) manned orbital flight was postponed at T-minus 29 minutes due to weather conditions.
NASA Historical Office, *Aerospace Chronology*, Jan. 1962.

January 30

The Mercury-Atlas 6 (MA-6) mission was postponed because of technical difficulties with the launch vehicle.
NASA Historical Office, *Aerospace Chronology*, Jan. 1962.

January (during the month)

Three potential recovery areas were recommended for the Mercury extended range or 1-day mission. These were: Grand Turk, Midway Island, and the Japanese-Philippine Island area.
NASA-MSC Report, *Project Mercury [Quarterly] Status Report No. 13 for Period Ending January 31, 1962.*

Modifications were started in order to use the New York-Bermuda submarine cable for the transmission of high speed radar data from the Bermuda network site to the Goddard Space Flight Center computers.

>NASA-MSC Report, *Project Mercury [Quarterly] Status Report No. 13 for Period Ending January 31, 1962.*

Twenty spacecraft aerial drop tests were planned for the Mercury extended range or 1-day mission. One of the prime objectives was to determine if the 63-foot ringsail main recovery parachute met all Mercury mission weight requirements. Tests were scheduled to be conducted at El Centro, California, and all tests would be land drops. This test program was designated Project Reef.

>NASA-MSC Report, *Project Mercury [Quarterly] Status Report No. 13 for Period Ending January 31, 1962.*

February 1

NASA Headquarters announced that the Mercury-Atlas 6 (MA-6) manned orbital mission would be scheduled no earlier than February 13, 1962, and that repair of the Atlas launch vehicle fuel tank leak would be completed well before that time.
>NASA Historical Office, *Aerospace Chronology*, Feb. 1962.

February 14

Unfavorable weather conditions caused the Mercury-Atlas 6 (MA-6) manned orbital mission to be postponed.
>NASA Historical Office, *Aerospace Chronology*, Feb. 1962.

February 16

Walter C. Williams, Project Mercury Operations Director, announced that because of weather conditions February 20, 1962, would be the earliest date that the Mercury-Atlas 6 mission could be launched.
>NASA Historical Office, *Aerospace Chronology*, Feb. 1962.

February 20

Mercury-Atlas 6 (MA-6) was launched from Cape Canaveral with astronaut John Glenn as pilot. (See fig. 60.) The Friendship 7 spacecraft covered its three-orbit flight in 4 hours 55 minutes, and 23 seconds. Some 60 million persons viewed astronaut Glenn's launch on live television. During the flight two major problems were encountered: (1) a yaw attitude control jet apparently clogged, forcing the astronaut to abandon the automatic control system for the manual-electrical fly-by-wire system and the manual-mechanical system; and (2) a faulty switch in the heat shield circuit indicated that the clamp holding the shield had been prematurely released - a signal later found to be false. During reentry, however, the retropack was not jettisoned but retained as a safety measure to hold the heat shield in place in the event it had loosened. The spacecraft landed in the Atlantic Ocean about 800 miles southeast of Bermuda and was recovered by the USS Noa after being in the water for 21 minutes. With the success of Mercury-Atlas 6 (MA-6) the basic objectives of Project Mercury had been reached - a man put into earth orbit, his reactions to space environment observed, and his safe return to earth to a point where he could be readily found. Prior to the flight, there was concern about the psychological effects of prolonged weightlessness. To the contrary, there were no debilitating or harmful effects, the astronaut found the zero g conditions very handy in performing his tasks, and felt exhilarated during his 4.5 hours of weightlessness. One of the interesting sidelights of the Glenn flight was his report of "fire flies" when he entered the sunrise portion of an orbit. For some time this phenomenon remained a space mystery, until Scott Carpenter accidentally tapped the spacecraft wall with his hand, releasing many of the so-called "fire flies." The source was determined to be frost from the reaction control jets.

> MSC-NASA Report, *Results of the First United States Manned Orbital Flight*, Feb. 20, 1962.

February 21

A metal fragment, identified by numbers stamped on it as a part of the Atlas that boosted Mercury-Atlas 6 (MA-6) into orbit, landed on a farm in South Africa after about 8 hours in orbit.

> NASA Historical Office, *Aerospace Science and Technology: A Chronology for 1962*, March 1962.

February 23

In a ceremony at Cape Canaveral, President John F. Kennedy awarded the NASA Distinguished Service Medal to John Glenn and Robert R. Gilruth.

> MSC Booklet, *Astronaut John H. Glenn, Jr., Friendship 7*, February 20, 1962.

February 25

Factory roll-out inspection of Atlas launch vehicle 107-D, designated for the Mercury-Atlas 7 (MA-7) manned orbital mission, was conducted at Convair.

> NASA-MSC Report, *Project Mercury [Quarterly] Status Report No. 14 for Period Ending April 30, 1962*

Figure 60. Mercury-Atlas 6: First manned (Glenn) orbital flight.

February 26

John Glenn Day in Washington, D.C., featured the reception of the astronaut at the White House, a parade, and his address to joint session of Congress.
NASA Historical Office, *Aerospace Science and Technology: A Chronology for 1962,* Feb. 1962.

March 1

An estimated 4 million people lined the streets of New York City for "John Glenn Day." Mayor Robert Wagner presented Glenn and Robert R. Gilruth the city's Medal of Honor.

NASA Historical Office, *Aerospace Science and Technology: A Chronology for 1962*, March 1962.

McDonnell submitted Mercury Report No. 8140, entitled "Contractor Furnished Equipment Status Report," showing the status of component qualification tests.

NASA-MSC Report, *Project Mercury [Quarterly] Status Report No. 14 for Period Ending April 30, 1962.*

March 2

The Mercury astronauts were guests of the United Nations, and John Glenn acted as spokesman during an informal reception given by Acting Secretary General U Thant.

NASA Historical Office, *Aerospace Science and Technology: A Chronology for 1962*, March 1962.

March 4-5

Scott Carpenter and Walter Schirra, designated (but not publicly) as pilot and backup pilot, respectively, for the Mercury-Atlas 7 (MA-7) manned orbital mission, underwent water-egress exercises. Several side-hatch egresses were made in conjunction with helicopter pickups.

NASA-MSC Report, *Project Mercury [Quarterly] Status Report No. 14 for Period Ending April 30, 1962.*

March 6

Atlas launch vehicle 107-D was delivered to Cape Canaveral for the Mercury-Atlas 7 (MA-7) mission.

Data supplied by Ken Vogel, Mercury Project Office, MSC.

March 7

The first Orbiting Solar Observatory (OSO) performed remarkably well in conducting the thirteen different experiments for which it was programmed. Especially relevant to manned space flight were its measurements of solar radiation in high frequency ranges, of cosmic dust effects, and of the thermal properties of spacecraft surface materials.

Goddard Space Flight Center chart: Satellites and Space Probe Projects as of July 1962.

March 9

John Glenn became the third man to be presented with Astronaut Wings in a ceremony at the Pentagon.

NASA Historical Office, *Aerospace Science and Technology: A Chronology for 1962*, March 1962.

March 12

During the period of the move of the Manned Spacecraft Center from Langley Field to Houston, Texas, primary Mercury operational activities remained at Langley to prevent any disruptions in the Mercury operational program.
> Robert R. Gilruth, MSC No. 21 2-1, subject: Relocation of Manned Spacecraft Center Headquarters, Feb. 26, 1962.

March 15

NASA Headquarters publicly announced that Scott Carpenter would pilot the Mercury-Atlas 7 (MA-7) manned orbital mission replacing Donald Slayton. The latter, formerly scheduled for the flight, was disqualified because of a minor erratic heart rate.
> NASA Historical Office, *Aerospace Science and Technology: A Chronology for 1962*, March 1962.

March 20

Spacecraft 19 was delivered to Cape Canaveral in the orbital-manned configuration, but this mission was canceled after the successful six-orbit flight of Schirra.
> Data supplied by Ken Vogel, Mercury Project Office, MSC.

March 22

Manned Spacecraft Center personnel briefed the Chief of Naval Operations on the Mercury-Atlas 7 (MA-7) flight and ensuing Mercury flights. This material was incorporated in a document entitled, "NASA Project Mercury Advance Recovery Requirements."
> NASA-MSC Report, *Project Mercury [Quarterly] Status Report No. 14 for Period Ending April 30, 1962*.

March (during the month)

The PERT (Program Evaluation and Review Technique) reporting system became operational on an experimental basis. The first PERT report on the Mercury 1-day mission schedule and cost analysis was issued by the Manned Spacecraft Center on April 26, 1962.
> NASA-MSC Report, *Project Mercury [Quarterly] Status Report No. 15 for Period Ending July 31, 1962*.

April 6

NASA sponsored a 1-day symposium in Washington on the results of the Mercury-Atlas 6 (MA-6) three-orbit flight of John Glenn. One of the items of particular interest was Glenn's "fire-flies," or luminous particles, and their possible origin.

NASA Historical Office, *Aerospace Science and Technology: A Chronology for 1962*, March 1962.

April 9

The National Geographic Society awarded the Hubbard Medal to John Glenn. This award has been made only 20 times since its origination in 1906. Glenn joined such recipients as Admiral Robert A. Peary, Charles A. Lindbergh, Ronald Amundsen, and Admiral Richard E. Byrd.

NASA Historical Office, *Aerospace Science and Technology: A Chronology for 1962*, April 1962.

April 15

Scott Carpenter and Walter Schirra, designated as pilot and backup pilot, respectively, for the Mercury-Atlas 7 (MA-7) manned orbital mission, underwent a water exercise training program to review procedures for boarding the life raft and the use of survival packs.

NASA-MSC Report, *Project Mercury [Quarterly] Status Report No. 14 for Period Ending April 30, 1962*.

April 19

NASA announced that the spacecraft, Friendship 7, used in the Mercury-Atlas 6 (MA-6) manned orbital mission would be lent to the United States Information Agency for a world tour, involving 20 stops and touching all continents. This tour was known as the "fourth orbit of Friendship 7." William Bland of the Mercury Project Office served as tour officer.

Letter, McDonnell Aircraft Corporation to Kenneth Kleinknecht, no subject, Sept. 10, 1962, with enclosure, "A Detailed Report on the Fourth Orbit of Friendship 7.".

April 30

Some 27 items of bite-size food were sampled and tested for possible inclusion in the Mercury space flights.

Activity Report, Life Systems Division, MSC, March 31, 1962 to April 30, 1962.

Swimmer training was started for the Mercury-Atlas 7 (MA-7) manned orbital mission recovery area. Instruction consisted of films, briefings, exercises in deploying the auxiliary flotation collar, and jumps from a helicopter.

NASA-MSC Report, *Project Mercury [Quarterly] Status Report No. 14 for Period Ending April 30, 1962*.

April (during the month)

Development of an advanced state-of-the-art pressure suit and helmet was started. This action was taken in preparation for the Mercury extended range or

1-day mission program. The objectives were aimed at improvements in unpressurized suit comfort, suit ventilation, pressure suit mobility, electrically heated helmet visor with additional light attenuation features, and the fabrication of a mechanical visor seal mechanism.
>NASA-MSC Report, *Project Mercury [Quarterly] Status Report No. 14 for Period Ending April 30, 1962.*

May 1

A gas analysis laboratory was installed in Hanger S at Cape Canaveral to analyze gases used in the Mercury spacecraft.
>Memo, G. Merritt Preston, Preflight Operations Division, to Director, MSC, subject: Monthly Activities Report No. 6, April 26, 1962.

May 4

A memorandum was issued on proposed experiments for inclusion in Mercury manned orbital flights. This action was in keeping with a statement made by Walter C. Williams before a University of Houston audience that the spacecraft would be used as a test bed for more ambitious space projects.
>Weekly Activity Report to the Office of the Director for Manned Space Flight, prepared by Management Analysis Office, MSC, April 29-May 5, 1962.

Scott Carpenter, designated as the primary pilot for the Mercury-Atlas 7 (MA-7) manned orbital flight completed a simulated MA-7 mission exercise.

>NASA Historical Office, *Aerospace Science and Technology: A Chronology for 1962*, May 1962.

May 7

NASA announced that the Mercury-Atlas 7 (MA-7) manned orbital flight would be delayed several days due to checkout problems with the Atlas launch vehicle.
>NASA Historical Office, *Aerospace Science and Technology: A Chronology for 1962*, May 1962.

May 15

Scott Carpenter, designated as the primary pilot of the Mercury-Atlas 7 (MA-7) manned orbital flight, flew a simulated mission with the spacecraft mated to the Atlas launch vehicle.
>Activity Report to the Office of the Director for Manned Space Flight, prepared by Management Analysis Division, MSC, May 13-19, 1962.

May 17

The Mercury-Atlas 7 (MA-7) manned orbital mission was postponed a second time because of necessary modifications to the altitude-sensing instrumentation in the parachute-deployment system.
    NASA Historical Office, *Aerospace Science and Technology: A Chronology for 1962*, May 1962.

May 19

A third postponement was made for the Mercury-Atlas 7 (MA-7) flight mission due to irregularities detected in the temperature control device on a heater in the Atlas flight control system.
    NASA Historical Office, *Aerospace Science and Technology: A Chronology for 1962*, May 1962.

May 24

Mercury-Atlas 7 (MA-7) was launched into earth orbit with astronaut Scott Carpenter as the pilot. The three-orbit flight of the spacecraft, designated Aurora 7, achieved all objectives. Only one critical component malfunction occurred during the mission - a random failure of the circuitry associated with the pitch horizon scanner, which provides a reference point to the attitude gyros. Also during the flight there was concern over the excessive fuel usage, a condition which resulted from extensive use of the high-thrust controls and the inadvertant use of two control systems simultaneously. To compensate, the spacecraft was allowed to drift for 77 minutes, in addition to the drifting already a part of the flight plan. The flight lasted for 4 hours and 56 minutes, and the spacecraft landed in the Atlantic Ocean 125 miles northeast of Puerto Rico, some 250 miles beyond the predicted impact point. The overshoot was traced to a 25 degree yaw error at the time the retrograde rockets were fired. Retrofire was about 3 seconds late, which accounts for about 20 miles of the overshoot. Computers at the Goddard Space Flight Center predicted the overshoot after the retrofire action. Carpenter was recovered by a helicopter and taken to the *USS Intrepid* after being in the water for 2 hours and 59 minutes. The astronaut did not incur any detrimental physical or biomedical effects. Two experiments were aboard the MA-7 spacecraft: one pertained to the behavior of liquid in a weightless state, and the other was a deployed balloon to measure drag and provide visibility data. The balloon failed to inflate properly (fig. 61), but the liquid reacted as had been anticipated (fig, 62). Carpenter also saw Glenn's "fire-flies."
    NASA, *Results of the Second United States Manned Orbital Space Flight, May 24, 1962*, SP-6.

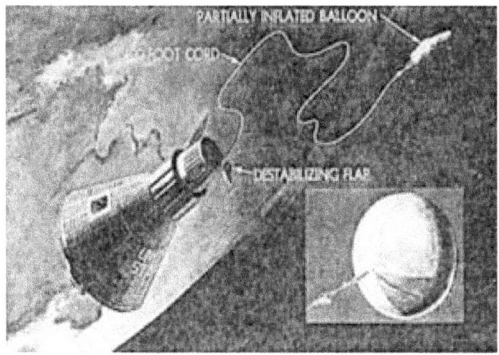
Figure 61. Balloon experiment.

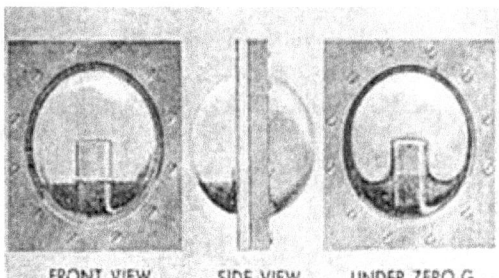
Figure 62. Zero-gravity experiment.

May 27

Scott Carpenter and Walter C. Williams were awarded the NASA Distinguished Service Medal by James Webb, NASA Administrator, in a ceremony at Cape Canaveral.
  MSC Booklet, *Astronaut M. Scott Carpenter, Aurora 7, May 24, 1962.*

May 28

Flight and ground tests disclosed that retrorocket heater blankets were unnecessary to the spacecraft, and this item was removed.
  MSC Highlights, prepared by Management Analysis Office, May 28, 1962.

For possible application purposes, and upon request, the Manned Spacecraft Center shipped Mercury-type survival kits to the Air Force for its X-20 Dyna Soar development program and to the Navy.

  MSC Highlights, prepared by Management Analysis Office, May 28, 1962.

May 31

Technical Report No. 138, entitled "Results of Project Mercury Ballistic and Orbital Chimpanzee Flights," was completed.
  Source as cited in text.

May (during the month)

Decision was made between April 29 and May 5, 1962, that leg supports would be removed from the Mercury couch. It had been determined that the heel and toe supports could be used as the sole supports for the lower leg. (See fig. 63.)
  Weekly Activity Report to Office of Director for Manned Space Flight, prepared by Management Analysis Division, April 29-May 5, 1962.

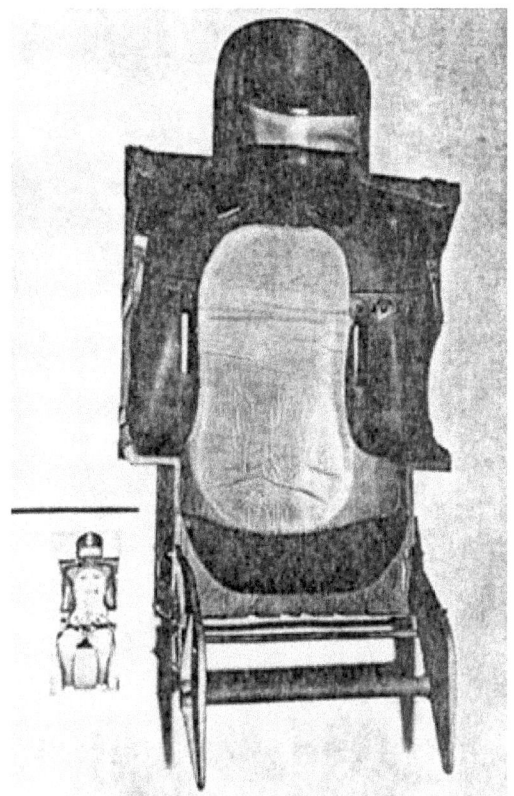

Figure 63. Astronaut couch modifications.

PART III (B)
Operational Phase of Project Mercury
June 1962 through June 12, 1963

1962

June 13

The Manned Spacecraft Center proposed a recoverable meteoroid experiment. According to the proposal, sheets of aluminum would be extended from the Mercury spacecraft and exposed to a meteoroid environment for a period of about 2 weeks. The sheets would then be retracted into the spacecraft for protection during reentry and recovery.
> Memo, John R. Davidson to Langley Associate Director, subject: Recoverable Meteoroid Penetration Experiment Using the Mercury Capsule as a Container, June 19, 1962.

June 25

Scott Carpenter was the fourth individual of Project Mercury to be presented Astronaut Wings by his respective service.
> Information supplied by Edwin M. Logan, Astronaut Activities Office, MSC.

June 26

Project Reef, an airdrop program to evaluate the Mercury 63-foot ringsail main parachute's capability to support the higher spacecraft weight for the extended range or 1-day mission was completed. Tests indicated that the parachute qualified to support the mission.
> NASA-MSC Report, *Project Mercury [Quarterly] Status Report No. 15 for Period Ending July 31, 1962*; Monthly Activities Report, Mercury Project Office, July, 1962.

June 27

D. Brainerd Holmes, NASA Director of Manned Space Flight, announced that the Mercury-Atlas 8 (MA-8) manned orbital mission would be programed for as many as six orbits. Walter Schirra was selected as the prime pilot with Gordon Cooper serving as backup.
> NASA Historical Office, *Aerospace Science and Technology: A Chronology for 1962*, June 1962.

NASA's Office of Advanced Research and Technology announced the appointment of Dr. Eugene B. Konecci as Director of Biotechnology and Human Research. Dr. Konecci will be responsible for directing research and development of future life support systems, advanced systems to protect man in the space environment, and research to assure man's performance capability in space.

    House Committee, *Astronautical and Aeronautical Events of 1962*, June 12, 1963.

June 28

The Manned Spacecraft Center requested that the Langley Research Center participate in acoustic tests of ablation materials on Mercury flight tests. Langley was to prepare several material specimens which would be tested for possible application in providing lightweight afterbody heat protection for Apollo class vehicles. Langley reported the results of its test preparation activities on September 21, 1962.
    Memo, Langley to Manned Spacecraft Center, subject: Transmittal of Data from Intense Noise Tests of Ablation Materials, Sept. 21, 1962.

June 29

Engineering was completed for the spacecraft reaction control system reserve fuel tank and related hardware in support of the Mercury extended range or 1-day mission.
    NASA-MSC Report, *Project Mercury [Quarterly] Status Report No. 15 for Period Ending July 31, 1962.*

June 30

Personnel strength of the Manned Spacecraft Center was 1,802.
    Statistics supplied by Katheryn Walker, Personnel Division, MSC.

July 1

Relocation of the Manned Spacecraft Center from Langley Field to Houston, Texas, was completed.
    MSC Fact Sheet No. 59, subject: NASA Manned Spacecraft Center Completes its Relocation to Temporary Houston Facilities, July 1, 1962.

July 8

Controversial Operation Dominic succeeded, after two previous attempts in June, in exploding a megaton-plus hydrogen device at more than 200-mile altitude over Johnston Island in the Pacific. Carried aloft by a Thor rocket and synchronized with the approach of a TRAAC satellite, this highest thermonuclear blast ever achieved was designed to test the influence of such an explosion on the Van Allen radiation belts. The sky above the Pacific Ocean from Wake Island to New Zealand was illuminated by the blast. Later observations by probes and satellites showed another artificial radiation belt to have been created by this series of nuclear tests.
    House Committee, *Astronautical and Aeronautical Events of 1962*, June 12, 1963.

July 9

NASA scientists concluded that the layer of haze reported by astronauts Glenn and Carpenter was a phenomenon called "airglow." Using a photometer, Carpenter was able to measure the layer as being 2 degrees wide. Airglow accounts for much of the illumination in the night sky.
> NASA Historical Office, *Aerospace Science and Technology: A Chronology for 1962*, July 1962.

July 11

NASA officials announced the basic decision for the manned lunar exploration program that Project Apollo shall proceed using the lunar orbit rendezvous as the prime mission mode. Based on more than a year of intensive study, this decision for the lunar orbit rendezvous (LOR), rather than for the alternative direct ascent or earth orbit rendezvous modes, enables immediate planning, research and development, procurement, and testing programs for the next phase of space exploration to proceed on a firm basis. (See 16 June 1961.)
> House Committee, *Astronautical and Aeronautical Events of 1962*, June 12, 1963.

July 13

Tests were conducted with a subject wearing a Mercury pressure suit in a modified Mercury spacecraft couch equipped with a B-70 (Valkyrie) harness. When this harness appeared to offer advantages over the existing Mercury harness, plans were made for further evaluation in spacecraft tests.
> MSC Life Systems Division, Weekly Activity Report, July 9-13, 1962.

July 21

President John F. Kennedy announced that Robert R. Gilruth, Director of Manned Spacecraft Center, would receive the President's Award for Distinguished Federal Civilian Service. This award was made for his successful accomplishment of "one of the most complex tasks ever presented to man in this country. . . the achievement of manned flight in orbit around the earth."
> NASA Historical Office, *Aerospace Science and Technology: A Chronology for 1962*, July 1962.

July 27

Atlas launch vehicle No. 113-D was inspected at Convair and accepted for the Mercury-Atlas 8 (MA-8) manned orbital mission.
> NASA-MSC Report, *Project Mercury [Quarterly] Status Report No. 15 for Period Ending July 31, 1962.*

August 6

The Friendship 7 spacecraft of the Mercury-Atlas 6 (MA-6) manned orbital mission (Glenn flight) was placed on display at the Century 21 Exhibition in Seattle, Washington. After this exhibition, the spacecraft was presented to the National Air Museum of the Smithsonian Institution, at formal presentation exercises on February 20, 1963.
Information by Robert Gordon, Exhibits and Displays, Public Affairs Office, MSC.

August 8

Spacecraft 9 (redesignated 9A) was phased into the Project Orbit program in preparation for the Mercury extended range or 1-day mission.
Actual testing began in September 1962. NASA-MSC Report, *Project Mercury [Quarterly] Status Report No. 15 for Period Ending July 31, 1962.*

Atlas launch vehicle 113-D was delivered to Cape Canaveral for the Mercury-Atlas 8 (MA-8) manned orbital mission.

Data supplied by Ken Vogel, Mercury Project Office, MSC.

August 10

NASA announced the appointment of Dr. Robert L. Barre as Scientist for Social, Economic, and Political Studies in the Office of Plans and Program Evaluation. Dr. Barre will be responsible for developing NASA's program of understanding, interpreting, and evaluating the social, economic, and political implications of NASA's long-range plans and accomplishments.
House Committee, *Astronautical and Aeronautical Events of 1962*, June 12, 1963.

August 11

A spacecraft reaction control system test was completed. Data compiled from this test was used to evaluate the thermal and thruster configuration of the Mercury extended range or 1-day mission spacecraft.
NASA-MSC Report, *Project Mercury [Quarterly] Status Report No. 16 for Period Ending October 31, 1962.*

August 11-12

U.S.S.R. launched VOSTOK III into orbit, piloted by Major Andrian G. Nikolayev. The next day, August 12, VOSTOK IV was launched into orbit piloted by Lt. Colonel Pavel R. Popovich, and near-rendezvous was achieved.
House Committee, *Astronautical and Aeronautical Events of 1962*, June 12, 1963.

August 15/

Navy swimmers, designated for the Mercury-Atlas 8 (MA-8) manned orbital mission recovery area, started refresher training at Pensacola, Florida. Instruction included installing the auxiliary flotation collar on a boilerplate spacecraft and briefings on assisting astronaut egress from the spacecraft.
> NASA-MSC Report, *Project Mercury [Quarterly] Status Report No. 15 for Period Ending July 31, 1962.*

August 21

A conference was held at the Rice Hotel, Houston, Texas, on the technical aspects of the Mercury-Atlas 7 (MA-7) manned orbital mission (Carpenter flight).
> Conference attended by author.

August (during the month)

The first edition of the map for the Mercury 1-day mission was published.
> USAF Aeronautical Chart and Information Center, 1st ed., Aug. 1962.

August-September

Negotiations were completed with McDonnell for spacecraft configuration changes to support the Mercury 1-day manned orbital mission. The design engineering inspection, when the necessary modifications were listed, was held on June 7, 1962.
> NASA-MSC Report, *Project Mercury [Quarterly] Status Report No. 15 for Period Ending July 31, 1962.*

September 7

The results of a joint study by the Atomic Energy Commission, the Department of Defense, and NASA concerning the possible harmful effects of the artificial radiation belt created by Operation Dominic on Project Mercury's flight MA-8 were announced. The study predicted that radiation on outside of capsule during astronaut Walter M. Schirra's six-orbit flight would be about 500 roentgens but that shielding, vehicle structures, and flight suit would reduce this dosage down to about 8 roentgens on the astronaut's skin. This exposure, well below the tolerance limits previously established, would not necessitate any change of plans for the MA-8 flight.
> House Committee, *Astronautical and Aeronautical Events of 1962*, June 12, 1963.

September 8

Atlas launch vehicle 113-D for the Mercury-Atlas 8 (MA-8) manned orbital mission was static-fired at Cape Canaveral. This test was conducted to check modifications that had been made to the booster for the purpose of smoother engine combustion.
> Mercury Project Office, Monthly Activity Report, Sept. 1962.

September 10

The Mercury-Atlas 8 (MA-8) manned orbital mission was postponed and rescheduled for September 28, 1962, to allow additional time for flight preparation.
>NASA Historical Office, *Aerospace Science and Technology: A Chronology for 1962*, Sept. 1962.

September 12

President John F. Kennedy visited the Manned Spacecraft Center and was shown exhibits including Mercury, Gemini, and Apollo spacecraft hardware.
>MSC Weekly Activity Report for the Office of the Director of Manned Space Flight, Sept. 2-8, 1962.

NASA announced it would launch a special satellite before the end of the year "to obtain information on possible effects of radiation on future satellites and to give the world's scientific community additional data on the artificial environment created by the radiation belt." The 100-pound satellite would be launched from Cape Canaveral into an elliptical orbit ranging from about 170-mile perigee to 10,350-mile apogee.
First "mystery" satellite in history of space exploration was launched, according to British magazine *Flight International*. The magazine said the satellite orbited at a height of 113 miles and reentered the earth's atmosphere 12 days later. The satellite was listed as belonging to the U.S. Air Force, but spokesman said this was a "scientific guess based on our assessment of previous satellite launchings." Launching was not confirmed, and no official U.S. listing included such a satellite.

>House Committee, *Astronautical and Aeronautical Events of 1962*, June 12, 1963.

September 17

Studies completed by the Navy Biophysics Branch of the Navy School of Aviation Medicine, Pensacola, Florida, disclosed that astronaut Glenn had received less than one-half the cosmic radiation dosage expected during his orbital flight. The Mercury-Atlas 6 (MA-6) spacecraft walls had served as excellent protection.
>NASA Historical Office, *Aerospace Science and Technology: A Chronology for 1962*, Sept. 1962.

September 18

Donald Slayton, one of the seven chosen for the astronaut training program, was designated Coordinator of Astronaut Activities at the Manned Spacecraft Center.
>Robert R. Gilruth, MSC Announcement No. 87 2-2, subject: Coordinator of Astronaut Activities, Sept. 18, 1962.

The NASA spacecraft test conductor and the Convair test conductor notified the interface committee chairman of the readiness-for-mate of the adapter-interface area of the Mercury-Atlas 8 (MA-8).

> Memo, O. L. Duggan to Associate Director, MSC, subject: MA-8 (Spacecraft 16) Interface Inspection, Sept. 18, 1962.

September 22

As an experiment, Walter Schirra planned to carry a special 2.5-pound hand camera aboard the Mercury-Atlas 8 (MA-8) spacecraft. (See fig. 64.) During the flight, the astronaut would attempt to arrive at techniques that could be applied to an advanced Nimbus weather satellite.
> NASA-MSC Report, *Project Mercury [Quarterly] Status Report No. 16 for Period Ending October 31, 1962.*

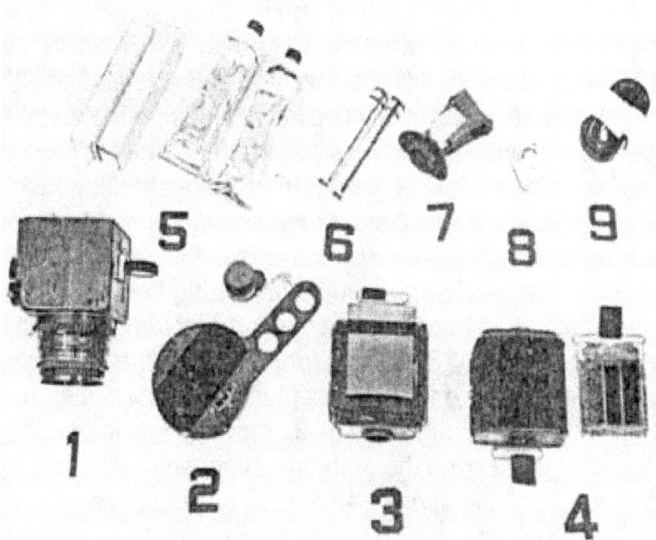

Figure 64. MA-8 ditty bag contents: 1. Camera (see text), 2. Photometer, 3. Color film magazine, 4. Film magazine, 5. Food containers, 6. Dosimeter, 7. Motion sickness container, 8. Exposure meter, 9. Camera shoulder strap.

September 28

Walter Schirra made a 6.5 hour simulated flight in the Mercury-Atlas 8 (MA-8) spacecraft. The worldwide tracking network of 21 ground stations and ships also participated in the exercise.
> NASA-MSC Report, *Project Mercury [Quarterly] Status Report No. 16 for Period Ending October 31, 1962.*

October 1

Tropical storm "Daisy" was studied by Mercury operations activities for its possible effects on the Mercury-Atlas 8 (MA-8) mission, but flight preparations continued.
> NASA Historical Office, *Aerospace Science and Technology: A Chronology for 1962*, Oct. 1962.

October 3

Mercury-Atlas 8 (MA-8), designated Sigma 7, was launched from Cape Canaveral with astronaut Walter Schirra as the pilot for a scheduled six-orbit flight. (See fig. 65.) Two major modifications had been made to the spacecraft to eliminate difficulties that had occurred during the Glenn and Carpenter flights. The reaction control system was modified to disarm the high-thrust jets and allow the use of low-thrust jets only in the manual operational mode to conserve fuel. A second modification involved the addition of two high frequency antennas mounted onto the retro package to assist and maintain spacecraft and ground communication throughout this flight. Schirra termed his six-orbit mission a "textbook flight." About the only difficulty experienced was attaining the correct pressure suit temperature adjustment. The astronaut became quite warm during the early orbits, but at a subsequent press conference he reported there had been many days at Cape Canaveral when he had been much hotter sitting under a tent on the beach. To study fuel conservation methods, a considerable amount of drifting was programed during the MA-8 mission. This included 118 minutes during the fourth and fifth orbits and 18 minutes during the third orbit. Since drift error was slight, attitude fuel consumption was no problem. At the start of the reentry operation there was a 78 percent supply in both the automatic and manual tanks, enabling Schirra to use the automatic mode during reentry. After a 9 hour and 13 minute orbital flight, the MA-8 landed 275 miles northeast of Midway Island, 9,000 yards from the prime recovery ship, the *USS Kearsarge*. Schirra stated that he and the spacecraft could have continued for much longer. The flight was the most successful to that time. Besides the camera experiment (September 22, 1962, entry), nine ablative material samples were laminated onto the cylindrical neck of the spacecraft, and radiation-sensitive emulsion packs were placed on each side of the astronaut's couch. As a note of unusual interest, the MA-8 launch was relayed via the Telstar satellite to television audiences in Western Europe.
> NASA, *Results of the Third United States Manned Orbital Space Flight, October 3, 1962*, SP-12.

Figure 65. Astronaut departs transfer van for Mercury-Atlas gantry.

October 5

Spacecraft 16, Sigma 7, was returned to Hanger S at Cape Canaveral for postflight work and inspection. It was planned to retain the Sigma 7 at Cape Canaveral for permanent display.
> NASA-MSC Report, *Project Mercury [Quarterly] Status Report No. 16 for Period Ending October 31, 1962*; Data supplied by Ken Vogel, Mercury Project Office, MSC; Message MSC-74 NASA-MSC-AMR Operations to Robert R. Gilruth, MSC, Oct. 29, 1962.

Dr. Charles A. Berry, Chief of Aerospace Medical Operations, Manned Spacecraft Center, reported that preliminary dosimeter readings indicated that astronaut Schirra had received a much smaller radiation dosage than expected.

NASA Historical Office, *Aerospace Science and Technology: A Chronology for 1962*, Oct. 1962.

A U.S. Air Force spokesman, Lt. Colonel Albert C. Trakowski, announced that special instruments on unidentified military test satellites had confirmed the danger that astronaut Walter M. Schirra, Jr., could have been killed if his MA-8 space flight had taken him above a 400-mile altitude. The artificial radiation belt, created by the U.S. high altitude nuclear test in July, sharply increases in density above 400-miles altitude at the geomagnetic equator and reaches peak intensities of 100 to 1,000 times normal levels at altitudes above 1,000 miles.

House Committee, *Astronautical and Aeronautical Events of 1962*, June 12, 1963.

October 7

The Mercury-Atlas 8 (MA-8) press conference was held at the Rice University, Houston, Texas. Astronaut Schirra expressed his belief that the spacecraft was ready for the 1-day mission, that he experienced absolutely no difficulties with his better than 9 hours of weightlessness, and that the flight was of the "textbook" variety.

Transcript of Press Conference (MA-8) held at Rice University, Oct. 8, 1962.

October 9

Spacecraft 20 was delivered to Cape Canaveral for the Mercury-Atlas 9 (MA-9) 1-day mission flight.

Data supplied by Ken Vogel, Mercury Project Office, MSC.

October 15

A high frequency direction finding system study was initiated. This study, covering a 12-month period, involved the development of high-frequency direction finding techniques to be applied in a network for locating spacecraft. The program was divided into a 5-month study and feasibility phase, followed by a 7-month program to provide operational tests of the procedures during actual Mercury flights or follow-on operations.

Letter, Kellogg Space Communications Laboratory to MSC, subject: Contract No. NAS 9-804, Oct. 22, 1962.

Walter Schirra was awarded the NASA Distinguished Service Medal by James Webb, NASA Administrator, for his six-orbit Mercury-Atlas 8 (MA-8) flight in a ceremony at his hometown, Oradell, New Jersey.

Information supplied by Ivan Ertel, Publi Affairs Office, MSC.

October 16

Walter Schirra became the fifth member of the Project Mercury team to receive Astronaut Wings.
> Information supplied by Ivan Ertel, Publi Affairs Office, MSC.

October 19

McDonnell reported that all spacecraft system tests had been completed for spacecraft 20, which was allocated for the Mercury-Atlas 9 (MA-9) 1-day orbital mission.
> McDonnell Report, subject: Model 133L- Project Mercury Spacecraft No. 20 Capsule Systems Tests, Oct. 19, 1962.

October 23

Major General Leighton Davis, Department of Defense representative for Project Mercury Support Operations, reported that support operation planning was underway for the Mercury 1-day mission.
> Letter, Air Force Missile Test Center to Robert S. McNamara, Secretary of Defense, subject: Department of Defense Support of Mercury-Atlas (MA-8), Oct. 23, 1962, with enclosures.

The Air Force Missile Test Center, Cape Canaveral, Florida, submitted a report to the Secretary of Defense summarizing Department of Defense support during the Mercury-Atlas 8 (MA-8) six-orbit flight mission.

> Letter, Air Force Missile Test Center to Robert S. McNamara, Secretary of Defense, subject: Department of Defense Support of Mercury-Atlas (MA-8), Oct. 23, 1962, with enclosures.

October 25

NASA Associate Administrator Robert C. Seamans, Jr., presented Outstanding Leadership Awards to Maxime A. Faget, Assistant Director for Engineering and Development, Manned Spacecraft Center, and George B. Graves, Jr., Assistant Director for Information and Control Systems. Also, at the NASA annual awards ceremony the Administrator, James E. Webb, presented Group Achievement Awards to four Manned Spacecraft Center activities: Assistant Directorate for Engineering and Development, Preflight Operations Division, Mercury Project Office, and Flight Operations Division.
> NASA Historical Office, *Aerospace Science and Technology: A Chronology for 1962*, Oct. 1962.

October 30

NASA announced realignment of functions within the office of Associate Administrator Robert C. Seamans, Jr. D. Brainerd Holmes assumed new duties as a Deputy Associate Administrator of Manned Space Flight. NASA field installations engaged primarily in manned space flight (Marshall Space Flight Center, MSC, and Launch Operations Center) would report to Holmes; installations engaged principally in other projects (Ames, Lewis Research Center, Langley Research Center, Goddard Space Flight Center, Jet Propulsion Laboratory, and Wallops Island) would report to Thomas F. Dixon, Deputy Associate Administrator for the past year. Previously, most field center directors had reported directly to Dr. Seamans on institutional matters beyond program and contractual administration.

    House Committee, *Astronautical and Aeronautical Events of 1962*, June 12, 1963.

November 1

Mercury Procedures Trainer No. 1, redesignated Mercury Simulator, was moved from Langley Field on July 23, 1962, and installed and readied for operations in a Manned Spacecraft Center building at Ellington Air Force Base, Houston, Texas.

    Activity Report, Flight Crew Operations Division, MSC, Aug. 20-28, 1962.

November 4

Enos, the 6-year-old chimpanzee who made a two-orbit flight around the earth aboard the Mercury-Atlas 5 (MA-5) spacecraft (November 29, 1961, entry) died at Holloman Air Force Base, New Mexico. The chimpanzee had been under night and day observation and treatment for 2 months before his death. He was afflicted with shigella dysentary, a type resistant to antibiotics, and this caused his death. Officials at the Air Medical Research Laboratory stated that his illness and death were in no way related to his orbital flight the year before.

    Information supplied by Aeromedical Research Center, Albuquerque, New Mexico, Nov. 6, 1962.

November 13

Gordon Cooper was named as the pilot for Mercury-Atlas 9 (MA-9) 1-day orbital mission slated for April 1963. Alan Shepard, pilot of Mercury-Redstone 3 (MR-3) was designated as backup pilot.

    NASA Hq. Release No. 52-245, subject: Cooper Named Pilot for MA-9 Flight, Nov. 14, 1962.

The B. F. Goodrich Company reported that it had successfully designed, fabricated, and tested a pivoted light attenuation tinted visor to be mounted on a government-issued Mercury helmet.

    Technical Report No. 62E-006, subject: Pivoted Light Attenuation Tinted Visor Helmet, submitted by B. F. Goodrich, Aerospace and Defense Products to MSC, Nov. 13, 1962.

November 16

The Manned Spacecraft Center presented the Department of Defense with recovery and network support requirements for Mercury-Atlas 9 (MA-9) 1-day manned orbital mission.
> Letter, MSC to Air Force Missile Test Center, subject: Project Mercury Preliminary Recovery and Network Requirements for MA-9, Nov. 16, 1962.

Mercury spacecraft 15A was delivered to Cape Canaveral for the Mercury-Atlas 10 (MA-10) orbital manned 1-day mission.

> Data supplied by Ken Vogel, Mercury Project Office, MSC (See also August 13, 1961, entry).

November 28

Mercury Simulator 2 was modified to the 1-day Mercury orbital configuration in preparation for the Mercury-Atlas 9 (MA-9) flight.
> Memo, Riley D. McCafferty to Mercury Project Office, subject: Mercury Flight Simulator Status Report, Nov. 28, 1962.

On this date, the McDonnell Aircraft Corporation reported that as of October 31, 1962, it had expended 4,231,021 man-hours in engineering; 478,926 man-hours in tooling; and 2,509,830 man-hours in production in support of Project Mercury.

> Letter, McDonnell to MSC, subject: Contract NAS 5-59, Mercury, Monthly Financial Report, Nov. 28, 1962.

Retrofire was reported to have initiated 2 seconds late during the Mercury-Atlas 8 (MA-8) mission. Because of this, the mechanics and tolerances of the Mercury orbital timing device were reviewed for the benefit of operational personnel, and the procedural sequence for Mercury retrofire initiation was outlined.

> Memo, J. D. Collier to John Hodge, Flight Operations Division, subject: Mercury Retrofire Initiation by the Spacecraft Satellite Clock, Nov. 28, 1962.

December 3-4

A pre-operational conference for the Mercury-Atlas 9 (MA-9) 1-day mission was held at Patrick Air Force Base, Florida, to review plans and the readiness status of the Department of Defense to support the flight. Operational experiences during the six-orbit Mercury-Atlas 8 (MA-8) mission were used as a planning guideline.
> Letter, Air Force Missile Test Center to MSC, *et al.*, subject: Minutes of Pre-Operational Conference for Project Mercury One-Day Mission (MA-9), Dec. 18, 1962.

December 4

Information was received from the NASA Inventions and Contributions activity that seven individuals, a majority of whom were still associated with the Manned Spacecraft Center, would receive monetary awards for inventions that were important in the development of Project Mercury. These were: Andre Meyer ($1,000) for the vehicle parachute and equipment jettison equipment; Maxime Faget and Andre Meyer (divided $1,500) for the emergency ejection device; Maxime Faget, William Bland, and Jack Heberlig (divided $2,000) for the survival couch; and Maxime Faget, Andre Meyer, Robert Chilton, Williard Blanchard, Alan Kehlet, Jerome Hammack, and Caldwell Johnson (divided $4,200) for the spacecraft design. Formal presentation of these awards was made on December 10, 1962.
>Letter, James A. Hootman, NASA Hq. to Dr. Robert R. Gilruth, MSC, subject: Monetary Awards for Project Mercury Inventors, Dec. 4, 1962.

December 7

The Massachusetts Institute of Technology Instrumentation Laboratory, charged with the development of the Apollo guidance and navigation system, was in the process of studying the earth's sunset limb to determine if it could be used as a reference for making observations during the mid-course phase of the mission. To obtain data for this study, the laboratory requested that photographic observations be made during the Mercury-Atlas 9 (MA-9) 1-day orbital mission. Photographic material from the Mercury-Atlas 7 (MA-7 - Carpenter flight) had been used in this study.
>Letter, Massachusetts Institute of Technology to MSC, Dec. 7, 1962.

December 14

Notice was received by the Manned Spacecraft Center from the NASA Office of International Programs that diplomatic clearance had been obtained for a survey trip to be conducted at the Changi Air Field, Singapore, in conjunction with Project Mercury contingency recovery operations. Also, the United Kingdom indicated informally that its protectorate, Aden, could be used for contingency recovery aircraft for the Mercury-Atlas 9 (MA-9) 1-day mission.
>Memo, Carl N. Jones, NASA Hq. to Christopher C. Kraft, MSC, subject: Use of Aden and Singapore for Contingency Recovery Purposes, Dec. 14, 1962.

December 15

Facilities at Woomera, Australia, a segment of the Mercury global network for telemetry reception and air-to-ground voice communications, was declared no longer required for Mercury flights.
>Memo, George M. Low to NASA Director of Network Operations and Facilities, subject: Fixture Requirements for Woomera Telemetry and Air-to-Ground Facilities, Dec. 15, 1962.

December 31

After reviewing Mercury-Atlas 9 (MA-9) recovery and network support requirements, the document covering the Department of Defense support of Project Mercury was forwarded to appropriate Department of Defense operational units for indication of their capability to fulfill requirements.
> Letter, Air Force Missile Test Center Hq. to COMDESFLOTFOUR, et al., subject: Department of Defense Support of Project Mercury Operations (MA-9), Dec. 31, 1962.

As of this date, the cumulative cost of the Mercury spacecraft design and development program with the McDonnell Aircraft Corporation, Contract NAS 5-59, had reached $135,764,042. During the tenure of this contract, thusfar, there had been 56 amendments and approximately 379 contract change proposals (CCP). At the end of the year, McDonnell had about 325 personnel in direct labor support of Project Mercury. Between March and May of 1960, the personnel complement had been slightly better than 1,600, representing a considerable rise from the 50 people McDonnell had assigned in January 1959 when study and contract negotiations were in progress. Peak assignments by month and by activity were as follows: Tooling - February 1960; Engineering - April 1960; and Production - June 1960.

> Letter, McDonnell to MSC, subject: Contract NAS 5-59, Mercury, Monthly Financial Report, Jan. 25, 1962, with enclosures.

December (during the month)

Three categories of experiments were proposed for the Mercury-Atlas 9 (MA-9) manned orbital mission: (1) space flight engineering and operations, (2) biomedical experiments, and (3) space science. The trailing balloon, similar to the MA-7 (Carpenter flight), was to be included. This balloon would be ejected, inflated, trailed, and jettisoned while in orbit. Another experiment was the installation of a self-contained flashing beacon installed on the retropackage, which would be initiated and ejected from the retropackage during orbital flight. And a geiger counter experiment was planned to determine radiation levels at varying orbital altitudes.
> Memo, Eugene M. Shoemaker, Chairman, Manned Space Science Planning Group to Director, Office of Manned Space Flight, Dec. 1962.

1963

January 3

Tentative plans were made by NASA to extend the Mercury-Atlas 9 (MA-9) flight from 18 to 22 orbits.

NASA Historical Office, *Astronautics and Aeronautics Chronology of Science and Technology in the Exploration of Space, January 1963.*

January 7

Final acceptance tests were conducted on the Mercury space flight simulator at Ellington Field, Texas. This equipment, formerly known as the procedures trainer, was originally installed at Langley Field and was moved from that area to Houston. Personnel of the Manned Spacecraft Center and the Farrand Optical Company conducted the acceptance tests.
> Letter, Farrand Optical Company, Inc., to NASA MSC, subject: Progress Report, Space Flight Simulator for Mercury Capsule, Jan. 22, 1963, with enclosures.

January 10-16

During this period, Mercury spacecraft No. 9A was cycled through Project Orbit Mission Runs 108, 108A, and 108B in the test facilities of the McDonnell Aircraft Corporation. These runs were scheduled for full-scale missions and proposed to demonstrate a 1-day mission capability. In other words, plans called for the operation of spacecraft systems according to the MA-9 flight plan, including the use of onboard supplies of electrical power, oxygen, coolant water, and hydrogen peroxide. Hard lines were used to simulate the astronaut control functions. Runs 108A and 108B were necessitated by an attempt to achieve the prescribed mission as cabin pressure difficulties forced a halt to the reaction control system thrust chamber operations portion of Run 108, although the other systems began to operate as programed. Later in 108 difficulties developed in the liquid nitrogen flow and leaks were suspected. Because of these thermal simulation problems, the test was stopped after 1 hour. Little improvement was recorded in Run 108A as leaks developed in the oxygen servicing line. In addition, cabin pressures were reduced to one psia, and attempts to repressurize were unsuccessful. The run was terminated. Despite the fact that Run 108B met with numerous problems - cabin pressure and suit temperature - a 40-hour and 30-minute test was completed.
> Letter, McDonnell Aircraft Corporation to NASA MSC, subject: Project Mercury, Model 133L, Project Orbit Spacecraft No. 9A T+ 3 Day Test Report, Category IV-I, Runs 108, 108A, 108B, Transmittal of, Jan. 25, 1963.

January 11

The Project Engineering Field Office (located at Cape Canaveral) of the Mercury Project Office reported on the number of changes made to spacecraft 20 (MA-9) as of that date after its receipt at Cape Canaveral from McDonnell in St. Louis. There were 17 specific changes, which follow: one to the reaction control system, one to the environmental control system, seven to the electrical and sequential systems, and eight to the console panels.
> Report, subject: Differences between Spacecraft 16 (MA-8) and Spacecraft 20 (MA-9) as of January 11, 1963, with Chart, subject: Summary of Spacecraft 20 Changes.

January 14

Mercury spacecraft 15A was redesignated 15B and allocated as a backup for the MA-9 mission. In the event Mercury-Atlas 10 (MA-10) were flown, 15B would be the prime spacecraft. Modifications were started immediately with respect to the hand controller rigging procedures, pitch and yaw control valves, and other technical changes.
> Memo, Wilbur Allaback to Kenneth Vogel, Engineering Operations, Mercury Project Office, subject: Project Mercury Spacecraft, Jan. 31, 1963.

The Manned Spacecraft Center presented the proposal to NASA Headquarters that the ground light visibility experiment of the Schirra flight (MA-8) be repeated for the Mercury-Atlas 9 (MA-9) mission. Objectives were to determine the capability of an astronaut visually to acquire a ground light of known intensity while in orbit and to evaluate the visibility of the light as seen from the spacecraft at varying distances from the light source. Possibly at some later date such lights could be used as a signal to provide spacecraft of advanced programs with an earth reference point. This experiment was integrated as a part of the MA-9 mission.

> Memo, Flight Control Operations Division to Mercury Project Manager, subject: Ground Light Visibility Experiment, Jan. 14, 1963.

January 17

Asked by a Congressional committee if NASA planned another Mercury flight after MA-9, Dr. Robert C. Seamans stated, in effect, that schedules for the original Mercury program and the 1-day orbital effort were presumed to be completed in fiscal year 1963. If sufficient test data were not accumulated in the MA-9 flight, backup launch vehicles and spacecraft were available to fulfill requirements.
> NASA Historical Office, *Astronautics and Aeronautics Chronology, January 1963.*

January 21

After reviewing the MA-9 spacecraft system and mission rules, the Simulations Section reported the drafting of a simulator training plan for the flight. Approximately 20 launch reentry missions were scheduled, plus variations of these missions as necessary. Instruction during the simulated orbital period consisted of attitude and fuel consumption studies, and from time to time fault insertions would be integrated to provide a complete range of activities covering all mission objectives. By the end of April 1963, the pilot and backup pilot had accumulated 50 hours in the simulators.
> Memo, Simulation Operations Section to Assistant Chief for Training, subject: MA-9 Astronaut Training, Jan. 21, 1963; NASA-MSC Report, *Project Mercury (MODM Project) [Quarterly] Status Report No. 18 for Period Ending April 30, 1963.*

January 22

McDonnell Aircraft Corporation reported to the Manned Spacecraft Center on a study conducted to ascertain temperature effects on the spacecraft as a result of white paint patch experiments. On both the MA-7 and MA-8 spacecraft, a 6-inch by 6-inch white patch was painted to compare shingle temperatures with an oxidized surface; the basic objective was to obtain a differential temperature measurement between the two surfaces, which were about 6 inches apart. Differences in spacecraft structural points prevented the tests from being conclusive, but the recorded temperatures during the flights were different enough to determine that the painted surfaces were cooler at points directly beneath the patch and on a corresponding point inside the spacecraft. According to McDonnell's analytical calculations, white painted spacecraft were advantageous for extended-range missions. However, McDonnell pointed out the necessity for further study, since one limited test was not conclusive.
    Letter, McDonnell Aircraft to NASA-MSC, subject: Contract NAS 5-59, Project Mercury, Models 133K and L, Results of McDonnell MMS-450 Type II White Paint on MA-7 and MA-8 Flights, Jan. 22, 1963.

January 26

Specialty assignments were announced by the Manned Spacecraft Center for its astronaut team: L. Gordon Cooper, Alan B. Shepard, pilot phases of Project Mercury; Virgil I. Grissom, Project Gemini; John H. Glenn, Project Apollo; M. Scott Carpenter, lunar excursion training; Walter M. Schirra, Gemini and Apollo operations and training; Donald K. Slayton, remained in duties assigned in September 1962 as Coordinator of Astronaut Activities. These assignments superseded those of July 1959. Assignments of the new flight-crew members selected on September 17, 1962, were as follows: Neil A. Armstrong, trainers and simulators; Frank Borman, boosters; Charles Conrad, cockpit layout and systems integration; James A. Lovell, recovery systems; James A. McDivitt, guidance and navigation; Elliott M. See, electrical, sequential, and mission planning; Thomas P. Stafford, communications, instrumentation, and range integration; Edward H. White, flight control systems; John W. Young, environmental control systems, personal and survival equipment.
    MSC Release 63-11, Jan. 26, 1963.

January 27

John A. Powers, Public Affairs Officer, Manned Spacecraft Center, told an audience of Texas Associated Press managing editors that Gordon Cooper's MA-9 flight might go as many as 22 orbits, lasting 34 hours.
    Associated Press Release, Jan. 28, 1963.

February 1

Kenneth S. Kleinknecht, Manager, Mercury Project Office, reported the cancellation of a peroxide expulsion experiment previously planned for the MA-9 mission. Kleinknecht noted the zodiacal light experiment would proceed and that the astronaut's gloves were being modified to facilitate camera operation.
    MSC Staff Meeting, Feb. 1, 1963.

February 5

Manned Spacecraft Center officials announced a delay of the MA-9 scheduled flight data due to electrical wiring problems in the Atlas launch vehicle control system.
    MSC Release 63-20, Feb. 5, 1963.

February 5-14

Personnel of the Manned Spacecraft Center visited the McDonnell plant in St. Louis to conduct a spacecraft status review. Units being inspected were spacecrafts 15B and 20. In addition, the status of the Gemini Simulator Instructor Console was assessed. With regard to the spacecraft inspection portion, a number of modifications had been made that would affect the simulator trainers. On spacecraft 15B, 15 modifications were made to the control panel and interior, including the relocation of the water separator lights, the addition of water spray and radiation experiment switches and a retropack battery switch. The exterior of the spacecraft underwent changes as well, involving such modifications as electrical connections and redesign of the fuel system for the longer mission. The reviewers found that spacecraft 20 conformed closely to the existing simulator configuration, so that modifications to the simulator were unnecessary.
    Memo, Simulator Operations Section to Assistant Chief for Training, subject: Mercury Spacecraft Status Review, March 5, 1963.

February 7

At a Development Engineering Inspection for the spacecraft 15B mockup, designated for the MA-10 mission, some 42 requests for alterations were listed.
    NASA-MSC Report, *Project Mercury (MODM Project) [Quarterly] Status Report No. 18 for Period Ending April 30, 1963.*

February 12

Objectives of the Mercury-Atlas 9 (MA-9) manned 1-day mission were published. This was the ninth flight of a production Mercury spacecraft to be boosted by an Atlas launch vehicle and the sixth manned United States space flight. According to plans, MA-9 would complete almost 22 orbits and be recovered approximately 70 nautical miles from Midway Island in the Pacific Ocean. Primary objectives of the flight were to evaluate the effects of the space environment on an astronaut after more than 1 but less than 2 days in orbit. During this period, close attention would be given to the astronaut's ability to function as a primary operating system of the spacecraft while in a sustained period of weightlessness. The capability of the spacecraft to perform over the extended period of time would be closely monitored. From postflight information, data would be available from the pilot and the spacecraft to ascertain, to a degree, the feasibility of space flights over a much greater period of time - Project Gemini, for example. In addition, the extended duration of the MA-9 mission provided a check on the effectiveness of the worldwide tracking network that could assist in determining the tracking requirement for the advanced manned space flight programs.

> NASA Project Mercury Working Paper No. 232, subject: Manned One-Day Mission Directive for Mercury-Atlas 9 (M-9, spacecraft 20), Feb. 12, 1963.

The Manned Spacecraft Center announced a mid-May flight for Mercury-Atlas 9 (MA-9). Originally scheduled for April, the launch date was delayed by a decision to rewire the Mercury-Atlas flight control system, as a result of the launch vehicle checkout at the plant inspection meeting.

> MSC Release 63-26, Feb. 12, 1963.

February 18-22

The McDonnell Aircraft Corporation reported to the Manned Spacecraft Center on the results of Project Orbit Run 109. This test run completed a 100-hour full-scale simulated mission, less the reaction control system operation, to demonstrate the 1-day mission capability of the Mercury spacecraft. Again, as in earlier runs, the MA-9/20 flight plan served as the guideline, including the use of onboard supplies of electrical power, oxygen, and coolant water, with hardline controls simulating astronaut functions. During the 2-hour prelaunch hold, a small leak was suspected in the secondary oxygen system, but at the end of the hold all systems indicated a "GO" condition and the simulated launch began. System equipment programing started and was recycled at the end of each 22 simulated orbits covering 33 mission hours. Test objectives were attained without any undue difficulty.

> Letter, McDonnell Aircraft Corporation to NASA-MSC, subject: Contract NAS 5-59, Project Mercury, Model 133L, Project Orbit Spacecraft No. 9A T+ 3 Day Test Report, Category IV-I, Run 109, Transmittal of, March 5, 1963.

February 20

Kenneth S. Kleinknecht, Manager, Mercury Project Office, commented on the first anniversary of the Glenn flight (MA-6) that 1,144.51 minutes of orbital space time had been logged by the three manned missions to date. These flights proved that man could perform in a space environment and was an important and integral part of the mission. In addition, the flights proved the design of the spacecraft to be technically sound. With the excellent cooperation extended by the Department of Defense, other government elements, industry, and academic institutions, a high level of confidence and experience was accrued for the coming Gemini and Apollo projects.

>MSC Release 63-20, Feb. 20, 1963.

The Smithsonian Institution received the Friendship 7 spacecraft (MA-6 Glenn flight) in a formal presentation ceremony from Dr. Hugh L. Dryden, the NASA Deputy Administrator. Astronaut John Glenn presented his flight suit, boots, gloves, and a small American flag that he carried on the mission.

>Information supplied by Albert M. Chop, Deputy Public Affairs Officer, Public Affairs Office, MSC, Feb. 21, 1963.

In announcing a realignment of the structure of the Office of Manned Space Flight, Director D. Brainerd Holmes named two new deputy directors and outlined a changed reporting structure. Dr. Joseph F. Shea was appointed Deputy Director for Systems, and George M. Low assumed duties as Deputy Director for Office of Manned Space Flight Programs. Reporting to Dr. Shea would be Director of Systems Studies, Dr. William A. Lee; Director of Systems Engineering, John A. Gautrand; and Director of Integration and Checkout, James E. Sloan. Reporting to Low would be Director of Launch Vehicles, Milton Rosen; Director of Space Medicine, Dr. Charles Roadman; and Director of Spacecraft and Flight Missions, presently vacant. Director of Administration, William E. Lilly, would provide administrative support in both major areas.

>NASA Release 63-32.

February 21

Gordon Cooper and Alan Shepard, pilot and backup pilot, respectively, for the Mercury-Atlas 9 (MA-9) mission, received a 1-day briefing on all experiments approved for the flight. Also at this time, all hardware and operational procedures to handle the experiments were established.

>NASA-MSC Report, *Project Mercury (MODM Project) [Quarterly] Status Report No. 18 for Period Ending April 30, 1963.*

The McDonnell Aircraft Corporation notified the Manned Spacecraft Center that the ultra high frequency transceivers were being prepared for the astronaut when in the survival raft. During tests of these components, an effective range of 5 to 10 miles had been anticipated, but the actual average range recorded by flyovers was 12 miles. Later, some faults were discovered in the flyover monitoring equipment, so that with adjustments the average range output was approximately 20 miles.

>Letter, McDonnell Aircraft Corporation to NASA-MSC, subject: Contract NAS 5-59, Project Mercury, Model 133L, UHF Transceiver Power Output, Feb. 21, 1963.

February 23

Manned Spacecraft Center checkout and special hardware installation at Cape Canaveral on spacecraft 20 were scheduled for completion as of this date. However, work tasks were extended for a 2-week period because of the deletion of certain experimental hardware - zero g experiment and new astronaut couch. In addition, some difficulties were experienced while testing the space reaction control system and environmental control system.
    Activity Report, MSC-Atlantic Missile Range Operations, January 27-February 23, 1963.

February (during the month)

The Air Force Aeronautical Chart and Information Center published the 22-orbit version of the worldwide Mercury tracking chart. The version of August 1962 covered 18 orbits.
    Mercury Orbit Chart MOC-6, 1st ed., USAF Aeronautical Chart and Information Center, Feb. 1963.

March 1

Spacecraft 9A, configured for manned 1-day mission requirements, completed Project Orbit Run 110. For this test, only the reaction control system was exercised; as a result of the run, several modifications were made involving pressurization and fuel systems.
    NASA-MSC Report, *Project Mercury (MODM Project) [Quarterly] Status Report No. 18 for Period Ending April 30, 1963.*

March 5

NASA Headquarters published a study on the ejection of an instrument package from an orbiting spacecraft. By properly selecting the ejection parameters, the package could be positioned to facilitate various observation experiments. From this experiment, if successful, the observation acuity, both visual and electrical, could be determined; this data would assist the rendezvous portion of the Gemini flights.
    NASA General Working Paper No. 10,005, subject: Parametric Study of Separation Distance and Velocity between a Spacecraft and an Ejected Object, March 5, 1963.

March 11

Based on a request from the Manned Spacecraft Center, McDonnell submitted a review of clearances between the Mercury spacecraft 15B retropack and the launch vehicle adapter during separation maneuvers. This review was prompted by the fact that additional batteries and a water tank had been installed on the sides of the retropack. According to the McDonnell study the clearance safety margin was quite adequate.
    Letter, McDonnell Aircraft Corporation to NASA MSC, subject: Contract NAS 5-59, Project Mercury, Model 133L, Retropack-Adapter Clearance Study, March 11, 1963.

March 15

Factory roll-out inspection of Atlas launch vehicle 130 was conducted at General Dynamics some 15 days later than planned. Delay was due to a re-work on the flight control wiring. After the launch vehicle passed inspection, shipment was made to Cape Canaveral on March 18, 1963, (see fig. 66) and the launch vehicle was erected on the pad on March 21, 1963.
NASA-MSC Report, *Project Mercury (MODM Project) [Quarterly] Status Report No. 18 for Period Ending April 30, 1963.*

Figure 66. Atlas launch vehicle 130-D (MA-9) undergoing inspection at Cape Canaveral.

March 19

The Manned Spacecraft Center received a slow-scan television camera system, fabricated by Lear Siegler, Incorporated, for integration with the Mercury-Atlas 9 (MA-9) mission. This equipment, weighing 8 pounds, could be focused on the pilot or used by the astronaut on other objects inside the spacecraft or to pick up exterior views. Ground support was installed at three locations - Mercury Control Center, the Canary Islands, and the Pacific Command Ship - to receive and transmit pictures of Cooper's flight. Transmission capabilities were one picture every 2 seconds.
MSC Release 63-52, March 19, 1963.

March 22

The National Rocket Club presented to John Glenn, pilot of America's first orbital manned space flight, the Robert H. Goddard trophy for 1963 for his achievement in assisting the advance of missile, rocket, and space flight programs.
MSC Release 63-54, March 20, 1963.

March 28

For the purpose of reviewing the MA-9 acceleration profile, pilot Gordon Cooper and backup pilot Alan Shepard received runs on the Johnsville centrifuge.
    NASA-MSC Report, *Project Mercury (MODM Project) [Quarterly] Status Report No. 18 for Period Ending April 30, 1963.*

April 5-6

Gordon Cooper and Alan Shepard, MA-9 pilot and backup pilot, visited the Morehead Planetarium in North Carolina to review the celestial sphere model, practice star navigation, and observe a simulation of the flashing light beacon (an experiment planned for the MA-9 mission).
    NASA-MSC Report, *Project Mercury (MODM Project) [Quarterly] Status Report No. 18 for Period Ending April 30, 1963.*

April 9

Langley Research Center personnel visited Cape Canaveral to provide assistance in preparing the tethered balloon experiment for the Mercury-Atlas 9 (MA-9) mission. This work involved installing force measuring beams, soldered at four terminals, to which the lead wires were fastened.
    Memo, Thomas Vranas to Associate Director, Langley Research Center, subject: Trip to Cape Canaveral to Rectify Difficulties in Strain Gage Instrumentation, April 25, 1963.

April 10-11

Full-scale recovery and egress training was conducted for Gordon Cooper and Alan Shepard in preparation for the MA-9 mission. During the exercise, egresses were effected from the spacecraft with subsequent helicopter pickup and dinghy boarding. The deployment and use of survival equipment were also practiced.
    NASA-MSC Report, *Project Mercury (MODM Project) [Quarterly] Status Report No. 18 for Period Ending April 30, 1963.*

April 15

The Manned Spacecraft Center published a detailed flight plan, and the assumption was made that the mission would be nominal, with any required changes being made by the flight director. Scheduled experiments, observations, and studies would be conducted in a manner that would not conflict with the operational requirements. Due to the extended duration of the flight, an 8-hour sleep period was programed, with a 2-hour option factor as to when the astronaut would begin his rest period. This time came well within the middle phases of the planned flight and would allow the astronaut ample opportunity to be in an alert state before retro-sequence. In addition to the general guidelines, the astronaut had practically a minute-to-minute series of tasks to accomplish.
    MA-9/20 Flight Plan, prepared by Spacecraft Operations Branch, Flight Crew Operations Division, MSC, April 15, 1963.

April 16-17

An MA-9 mission briefing was conducted for the astronauts and Mercury support personnel. Subjects under discussion included recovery procedures, network communications, spacecraft systems, flight plan activities, and mission rules.
    NASA-MSC Report, *Project Mercury (MODM Project) [Quarterly] Status Report No. 18 for Period Ending April 30, 1963.*

April 20

The final water condensate tank was installed in spacecraft 20 for the MA-9 mission. In all, the system consisted of a 4-pound, built-in tank, a 3.6 pound auxiliary tank located under the couch head, and six 1-pound auxiliary plastic containers. The total capacity for condensate water storage was 13.6 pounds. In operation, the astronaut hand-pumped the fluid to the 3.6 pound tank to avoid spilling moisture inside the cabin from the built-in tank. Then the 1-pound containers were available.
    Report, subject: Project Mercury Weekly Report 29, Spacecraft 20, April 21-27, 1963.

April 22

Spacecraft 20 was moved from Hanger S at Cape Canaveral to Complex 14 and mated to Atlas launch vehicle 130-D in preparation for the Mercury-Atlas 9 (MA-9) mission. The first simulated flight test was begun immediately.
    Report, subject: Project Mercury Weekly Report 29, Spacecraft 20, April 21-27, 1963.

The Bendix Corporation reported to the Manned Spacecraft Center that it had completed the design and fabrication of an air lock for the Mercury spacecraft. This component was designed to collect micrometeorites during orbital flight. Actually the air lock could accommodate a wide variety of experiments, such as ejecting objects into space and into reentry trajectories, and exposing objects to a space environment for observation and retrieval for later study. Because of the modular construction, the air lock could be adapted to the Gemini and Apollo spacecraft.

    Letter, Bendix Corporation to NASA-MSC, subject: Bendix Utica Air Lock for Mercury Spacecraft, April 22, 1963, with enclosures.

Scott Carpenter told an audience at the American Institute of Aeronautics and Astronautics' Second Manned Space Flight Meeting in Dallas, Texas, that the Mercury program would culminate with the 1-day mission of Gordon Cooper.

    Paper, subject: Flight Experiences in the Mercury Program, presented by M. Scott Carpenter, NASA MSC at the AIAA 2nd Manned Space Flight Meeting, Dallas, Texas, April 22, 1963.

After spacecraft 20 was mated to Atlas launch vehicle 130-D, a prelaunch electrical mate and abort test and a joint flight compatibility test were made.

During the latter, some difficulty developed in the flight control gyro canisters, causing replacement of the components; a rerun of this portion of the test was scheduled for May 1, 1963.

> NASA-MSC Report, *Project Mercury (MODM Project) [Quarterly] Status Report No. 18 for Period Ending April 30, 1963.*

April 30

As of this date, a number of improvements had been made to the Mercury pressure suit for the Mercury-Atlas 9 (MA-9) flight. (See fig. 67.) These included a mechanical seal for the helmet, new gloves with an improved inner-liner and link netting between the inner and outer fabrics at the wrist, and an increased mobility torso section. The MA-9 boots were integrated with the suit to provide additional comfort for the longer mission, to reduce weight, and to provide an easier and shorter donning time. Another change relocated the life vest from the center of the chest to a pocket on the lower left leg. This modification removed the bulkiness from the front of the suit and provided for more comfort during the flight. These are but a few of the changes.

> NASA-MSC Report, *Project Mercury (MODM Project) [Quarterly] Status Report No. 18 for Period Ending April 30, 1963*; information supplied by James McBarron, Crew Systems Division, MSC, May 13, 1963.

May 12

Dr. Charles A. Berry, Chief, Aerospace Medical Operations Office, Manned Spacecraft Center, pronounced Gordon Cooper in excellent mental and physical condition for the upcoming Mercury-Atlas 9 (MA-9) mission.

> MA-9 Advisory, Mercury Atlantic News Center, May 12, 1963.

May 12-19

Some 1,020 reporters, commentators, technicians, and others of the news media from the U.S. and several foreign countries gathered at Cape Canaveral, with another 130 at the NASA News Center in Hawaii, to cover the Mercury-Atlas 9 (MA-9) mission. Over the course of these days at Cape Canaveral, Western Union estimated that approximately 600,000 words of copy were filed, of which 140,000 were transmitted to European media. This does not include stories phoned in by reporters nor copy filed from the Pacific News Center, or for radio and TV coverage. During the 546,167-statute-mile flight, television audiences could see the astronaut or views inside and outside the spacecraft from time to time. Approximately 1 hour and 58 minutes were programed. Visual coverage was relayed to Europe via satellites.

> Observed by author; Western Union statistics supplied by John J. Peterson, Manager, Mercury Atlantic News Center, May 19, 1963.

Figure 67. Flight pressure suit of astronaut L. Gordon Cooper used in MA-9, 22-orbit mission. Compare with flight pressure suit of astronaut Alan Shepard in MR-3 suborbital flight in figure 37.

May 14

An attempt was made to launch Mercury-Atlas 9 (MA-9), but difficulty developed in the fuel pump of the diesel engine used to pull the gantry away from the launch vehicle. This involved a delay of approximately 129 minutes after the countdown had reached T-60 minutes. After these repairs were effected, failure at the Bermuda tracking station of a computer converter, important in the orbital insertion decision, forced the mission to be canceled at T-13 minutes. At 6:00 p.m. EDT, Walter C. Williams reported that the Bermuda equipment had been repaired, and the mission was rescheduled for May 15, 1963.
  Observed by the author.

May 15-16

Scheduled for a 22-orbit mission, Mercury-Atlas 9 (MA-9), designated Faith 7, was launched from Cape Canaveral at 8:04 a.m. EDT, with astronaut L. Gordon Cooper as the pilot. (See fig. 68.) Cooper entered the spacecraft at 5:33 a.m. the morning of May 15, and it was announced over Mercury Control by Lt. Colonel John A. Powers that "barring unforeseen technical difficulties the launch would take place at 8:00 a.m. EDT " As a note of interest, Cooper reported that he took a brief nap while awaiting the launch. The countdown progressed without incident until T-11 minutes and 30 seconds when some difficulty developed in the guidance equipment and a brief hold was called. Later, a momentary hold was called at T-19 seconds to determine whether the systems went into automatic sequencing, which occurred as planned. The liftoff was excellent, and visual tracking could be made for about 2 minutes through a cloudless sky. The weather was considerably clearer than on the day before. The Faith 7 flight sequencing - booster engine cut off, escape tower jettison, sustainer engine cut off - operated perfectly and the spacecraft was inserted into orbit at 8:09 a.m. EDT at a speed that was described as almost unbelievably correct. The perigee of the flight was about 100.2 statute miles, the apogee was 165.9, and Faith 7 attained a maximum orbital speed of 17,546.6 miles per hour. During the early part of the flight, Cooper was busily engaged in adjusting his suit and cabin temperatures, which were announced as 92 degrees and 109 degrees F, respectively, well within the tolerable range. By the second orbit, temperature conditions were quite comfortable, so much so, in fact, that the astronaut took a short nap. During the third orbit, Cooper deployed the flashing light experiment successfully and reported that he was able to see the flashing beacon on the night side of the fourth orbit. Thus Cooper became the first man to launch a satellite (the beacon) while in orbital flight. Another experiment was attempted after 9 hours in flight, during the sixth orbit, when Cooper tried to deploy a balloon but this attempt met with failure. A second deployment effort met with the same results. During this same orbit (sixth), the astronaut reported that he saw a ground light in South Africa and the town from which it emanated. This was an experiment to evaluate an astronaut's capability to observe a light of known intensity and to relate its possible applications to the Gemini and Apollo programs, especially as it pertained to the landing phases. After the beginning of the eighth orbit, Faith 7 entered a period of drifting flight - that is, the astronaut did not exercise his controls - and this drifting condition was programed through the fifteenth orbit. Some drifting flight had already been accomplished. Since the astronaut's sleep-option period was scheduled for this flight phase, Cooper advised the telemetry command ship, *Rose Knot Victor* off the coast of Chile, just before the ninth orbit that he planned to begin his rest period. The astronaut contacted John Glenn off the coast of Japan while in the ninth orbit, but lapsed into sleep shortly after entering the 10th orbit. During his sleep period, suit temperature rose slightly, and the astronaut roused, reset the control, and resumed his rest. Cooper contacted Muchea, Australia, during the 14th orbit after a restful night of drifting flight and resumed his work program. He reported just before entering the 17th orbit that he was attempting to photograph the zodiacal light. While in the 19th orbit, the first spacecraft malfunction of concern occurred when the .05g light appeared on the instrument panel as Cooper was adjusting the cabin light dimmer switch. The light, sensitive to gravity, normally lights during reentry. The flight director instructed Cooper to power up his attitude control system and to relay information on attitude indications reception on his gyros. All telemetry data implied that there had been no orbital decay and that the speed of Faith 7 was correct. The obvious conclusion was that the light was erroneous. Inspectors were later of the opinion that water in the system had caused a short circuit and had tripped a relay, causing the light to appear. Because of this condition, the flight director believed that certain portions of the automatic system would not work during reentry, and the astronaut was advised to reenter in the manual mode, becoming the first American to use this method exclusively for reentry. During the reentry operation, Cooper fired the retrorockets manually, by pushing a button for the first of three rockets to start the sequence. He attained the proper reentry attitude by using his observation window scribe marks to give proper reference with the horizon and to determine if he were rolling. John Glenn, aboard the command ship off the Japanese coast provided the countdown for the retrosequence and also advised Cooper when to jettison the retropack. The main chute deployed at 11,000 feet. Faith 7 landed 7,000 yards from the prime recovery ship, the carrier USS Kearsarge, after a 34-hour, 19-minute, and 49-second space flight. Cooper did not egress from the spacecraft until he was hoisted aboard the carrier. The mission was an unqualified success. During the flight the use of consumables - electrical power, oxygen, and attitude fuel - ran considerably below the flight plan. On the 15th orbit 75 percent of the primary supply of oxygen remained, and the reserve supply was untouched. The unusual low consumption rate of all supplies prompted teasing by the Faith 7 communicators. They called the astronaut a "miser" and requested that he "stop holding his breath."

MA-9 Transcript, May 15, 1963.

Figure 68. Astronaut L. Gordon Cooper prepares for insertion in Faith 7 (MA-9) atop Mercury-Atlas gantry for 22-orbit flight.

May 15

As of this date, the number of contractor personnel at Cape Canaveral directly involved in supporting Project Mercury were as follows: McDonnell, 251 persons for Contract NAS 5-59 and 23 persons for spacecraft 15B (MA-10 work); Federal Electric Corporation, 8. This report corresponded with the launch date of astronaut Gordon Cooper in the Mercury-Atlas 9 (MA-9).
    Memo, Harold G. Collins, Contracting Officer, to Director for Mission Requirements, subject: MSC/Cape Monthly Report on Contractor Personnel Headcount, May 17, 1963.

May 19

On a national televised press conference, emanating from Cocoa Beach, Florida, astronaut Gordon Cooper reviewed his experiences aboard the Faith 7 during the Mercury-Atlas 9 (MA-9) mission. Cooper, in his discussion, proceeded systematically throughout the mission from countdown through recovery. He opened his comments by complimenting Calvin Fowler of General Dynamics for his fine job on the console during the Atlas launching. During the flight, he reported that he saw the haze layer formerly mentioned by Schirra during the Sigma 7 flight (MA-8) and John Glenn's "fireflies" (MA-6). As for the sleep portion, Cooper felt he had answered with finality the question of whether sleep was possible in space flight. He also mentioned that he had to anchor his thumbs to the helmet restraint strap to prevent his arms from floating, which might accidentally trip a switch. Probably the most astonishing feature was his ability visually to distinguish objects on the earth. He spoke of seeing an African town where the flashing light experiment was conducted; he saw several Australian cities, including the large oil refineries at Perth; he saw wisps of smoke from rural houses on the Asiatic Continent; and he mentioned seeing Miami Beach, Florida, and the Clear Lake area near Houston. With reference to particular problems while in flight, the astronaut told of the difficulties he experienced with the condensate water pumping system. During the conference, when Dr. Robert C. Seamans was asked about the possibilities of a Mercury-Atlas 10 (MA-10) flight, he replied that "It is quite unlikely."
    MA-9 Press Conference, May 19, 1963.

May 21

In a White House ceremony, President John F. Kennedy presented astronaut Gordon Cooper with the NASA Distinguished Service Medal. Other members of the Mercury operations team receiving medals for outstanding leadership were as follows: G. Merritt Preston, Manager of Project Mercury Operations at Cape Canaveral; Floyd L Thompson, Langley Research Center; Kenneth S. Kleinknecht, Manager of the Mercury Project Office; Christopher C. Kraft, Director of Flight Operations Division, Manned Spacecraft Center; and Major General Leighton I. Davis, Commander, Air Force Missile Test Center.
    NASA Historical Office, *Astronautics and Aeronautics, May 1963.*

May 22

President Kennedy at a regular press conference responded to a question regarding the desirability of another Mercury flight by saying that NASA should and would make that final judgment.
    Transcript, New York *Times*, May 23, 1963, p. 18.

May 24

William M. Bland, Deputy Manager, Mercury Project Office, told an audience at the Aerospace Writers' Association Convention at Dallas, Texas, that "contrary to common belief, the Mercury spacecraft consumables have never been stretched like a rubber band to their limit in performing any of the missions." He pointed out that consumables such as electrical power, coolant water, oxygen, and carbon dioxide absorption were always available with large safety margins at the close of the flights. For example, astronaut Walter Schirra had a 9-hour primary oxygen supply at the end of his flight.
    Paper, William M. Bland, Jr. and Lewis R. Fisher, Project Mercury Experience.

May 29

The Department of Defense submitted a summary of its support of the Mercury-Atlas 9 (MA-9) mission, with a notation that the department was prepared to provide support for the MA-10 launch. Other than the provision of the Atlas launch vehicle, the Department of Defense supplied the Air Force *Coastal Sentry Quebec*, positioned south of Japan to monitor and backup retrofire for orbits 6, 7, 21, and 22. In the southeast Pacific, the Atlantic Missile Range telemetry command ship *Rose Knot Victor* was positioned to provide command coverage for orbits 8 and 13. At a point between Cape Canaveral and Bermuda, the Atlantic Missile Range C-band radar ship *Twin Falls Victory* was stationed for reentry tracking, while the Navy's *Range Tracker* out of the Pacific Missile Range provided similar services in the Pacific. Other Department of Defense communications support included fixed island stations and aircraft from the several services. Rear Admiral Harold G. Bowen was in command of Task Force 140, positioned in the Atlantic Ocean in the event of recovery in that area. In addition, aircraft were available at strategic spots for sea recovery or recovery on the American or African Continents. In the Pacific, recovery Task Force 130, under the command of Rear Admiral C.A. Buchanan, was composed of one aircraft carrier and 10 destroyers. This force was augmented by aircraft in contingency recovery areas at Hickam; Midway Island; Kwajalein; Guam; Tachikawa, Japan; Naha, Okinawa; Clark Field, Philippines; Singapore; Perth, Australia; Townsville; Nandi; Johnston Island; and Tahiti. Pararescuemen were available at all points except Kwajalein. The Middle East recovery forces (Task Force 109) were under the direction of Rear Admiral B.J. Semes and consisted of a seaplane tender and two destroyers supported by aircraft out of Aden, Nairobe, Maritius, and Singapore for contingency recovery operations. For bioastronautic support, the Department of Defense deployed 78 medical

personnel, had 32 specialty team members on standby, committed 9 department hospitals and provided over 3,400 pounds of medical equipment. During the actual recovery, the spacecraft was sighted by the carrier USS Kearsarge (Task Force 130), and helicopters were deployed to circle the spacecraft during its final descent. Swimmers dropped from the helicopters to fix the flotation collar and retrieve the antenna fairing. Cooper remained in his spacecraft until he was hoisted aboard the carrier. A motor whaleboat towed the spacecraft alongside the ship.
> Letter, Major General L.I. Davis, Department of Defense Representative Project Mercury/Gemini Support Operations, to Hon. Robert S. McNamara, Secretary of Defense, subject: Department of Defense Support of Project Mercury Manned One-Day Mission (MA-9), May 29, 1963.

Astronaut Gordon Cooper became the sixth Mercury astronaut to be presented with Astronaut Wings by his respective service.

> NASA Historical Office, *Astronautics and Aeronautics, May 1963*.

June 6-7

Officials of the Manned Spacecraft Center made a presentation to NASA Administrator James E. Webb, outlining the benefits of continuing Project Mercury at least through the Mercury Atlas 10 (MA-10) mission. They thought that the spacecraft was capable of much longer missions and that much could be learned about the effects of space environment from a mission lasting several days. This information could be applied to the forthcoming Projects Gemini and Apollo and could be gained rather cheaply since the MA-10 launch vehicle and spacecraft were available and nearing a flight readiness status.
> MSC Weekly Activity Report, June 2-8, 1963.

June 8

In preparation for the Mercury-Atlas 10 (MA-10) mission, should the flight be approved by NASA Headquarters, several environmental control system changes were made in spacecraft 15B. Particularly involved were improvements in the hardware and flexibility of the urine and condensate systems. With regard to the condensate portion, Gordon Cooper, in his press conference, indicated that the system was not easy to operate during the flight of Faith 7 (MA-9).
> MSC Weekly Activity Report, June 2-8, 1963.

June 12

Testifying before the Senate Space Committee, James E. Webb, the NASA Administrator, said: "There will be no further Mercury shots . . ." He felt that the manned space flight energies and personnel should focus on the Gemini and Apollo programs. Thus, after a period of 4 years, 8 months, and 1 week, Project Mercury, America's first manned space flight program, came to a close.
> MSC, *Space News Roundup*, June 26, 1963.

Appendix 1 - Project Mercury History

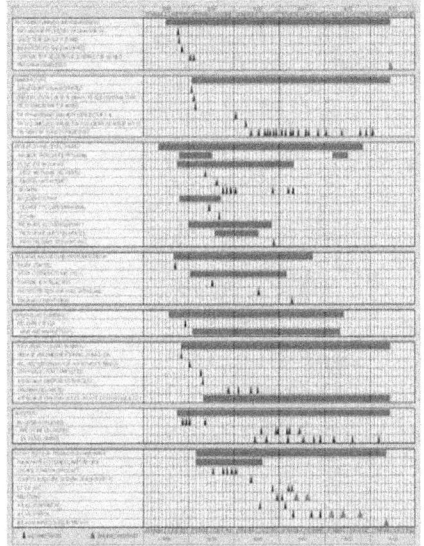

Appendix 2 - Project Mercury Test Objectives

MERCURY-LITTLE JOE MISSION AND TEST OBJECTIVES

Mission	Launch Date	Objectives
LJ-6	October 4, 1959	(a) To qualify the aerodynamics and structural integrity of the booster and the mechanical performance of the launcher. (b) To check the performance of the system for transmitting a command signal from the ground station, receiving it in the booster during flight, and setting of an explosive system at the head end of each main rocket motor in the booster.
LJ-1A	November 4, 1959	(a) To carry out a planned abort of the spacecraft from the booster at the maximum dynamic pressure anticipated during Mercury-Atlas exit flight. (b) To obtain added reliability data on the Mercury drogue and main parachute operation. (c) To study spacecraft impact behavior. (d) To gain further operational experience in recovery of a floating spacecraft, utilizing a surface vessel. (e) To obtain further experience and confidence in the operation of the booster command thrust termination system. (f) To recover escape motor and tower.
LJ-2 Primate aboard	December 4, 1959	(a) To carry out a planned escape of the spacecraft from the booster at high altitude (96,000 ft) just prior to main booster rocket motor burnout. (b) To ascertain spacecraft entry dynamics for an uncontrolled entry. (c) To check spacecraft dynamic stability on descent through the atmosphere without a drogue parachute. (d) To determine the physiological and psychological effects of acceleration and weightlessness on a small primate (rhesus monkey). (e) To obtain additional reliability data on the operation of the Mercury parachutes. (f) To obtain more data on Mercury spacecraft

		flotation characteristics in sea areas typical of those planned for use as recovery areas. (g) To obtain additional operational experience of spacecraft recovery by a surface vessel.
LJ-1B Primate aboard	January 21, 1960	(a) To check out the Mercury escape system concept and hardware at the maximum dynamic pressure anticipated during a Mercury-Atlas exit flight. (b) To determine the effects of simulated Atlas abort accelerations on a small primate (female rhesus monkey). (c) To obtain further reliability data on the Mercury spacecraft drogue and main chute operations. (d) To check out the operational effectiveness of spacecraft recovery by helicopter. (e) To recover the escape-system assembly (escape motor and tower) for a postflight examination in order to establish whether there had been any component malfunction or structure failure.
LJ-5A	March 18, 1961	(a) Demonstrate the structural integrity of the Mercury spacecraft and escape system during an escape initiated at the highest dynamic pressure that can be anticipated during an Atlas launch for orbital flight. (b) Demonstrate the performance of the spacecraft escape system, the sequential system, and the recovery system. (c) Determine the flight dynamic characteristics of the Mercury spacecraft in an escape maneuver. (d) Demonstrate the performance of a particular landing-bag configuration. (e) Establish the adequacy of the spacecraft recovery procedures. (f) Establish prelaunch checkout procedures for the functioning spacecraft systems. (g) Determine the effects of the flight profile on the spacecraft equipment and systems not otherwise required for the first-order test objectives.
LJ-5B	April 28, 1961	(a) Demonstrate the structural integrity of the Mercury spacecraft and escape system during an escape initiated at the highest dynamic pressure that can be anticipated during an Atlas launch for orbital flight. (b) Demonstrate the performance of the spacecraft

		escape system, the sequential system, landing system, and the recovery system. (c) Determine the flight dynamic characteristics of the Mercury spacecraft in an escape maneuver. (d) Establish the adequacy of the spacecraft recovery procedures. (e) Establish prelaunch checkout procedures for the functioning spacecraft systems. (f) Determine the effects of the flight profile on the spacecraft equipment and systems not otherwise required for first-order test objectives.

MERCURY BEACH ABORT TEST OBJECTIVES

Mission	Launch Date	Objectives
Beach Abort (Boilerplate spacecraft)	May 9, 1960	(a) Demonstrate capability of escape system, landing system, and postlanding equipment during an off-the-pad abort. (b) Demonstrate structural integrity of escape configuration during an off-the-pad abort. (c) Provide time history data for the following parameters: (1) altitude, (2) range, (3) velocity, (4) pitch, roll and yaw angles, (5) pitch, roll and yaw rates, (6) pitch, roll and yaw accelerations, (7) impact accelerations, and (8) sequence of events. (d) Obtain operational experience for checkout, launch and recovery teams. (e) Determine the effects of off-the-pad escape and landing conditions upon the spacecraft telemetry, instrumentation and communications systems. (f) Provide time history data for the following parameters: (1) indicated pressure altitude, (2) outside skin temperature, (3) inside skin temperature, (4) cabin air temperature, (5) noise level, and (6) vibration.

MERCURY-REDSTONE MISSIONS AND TEST OBJECTIVES

Mission	Launch Date	Objectives
MR-1A Unmanned	December 19, 1960	(a) Qualify the spacecraft-booster combination for the Mercury-Redstone mission which includes attaining a Mach number of approximately 6.0 during powered flight, a period of weightlessness of about 5 minutes, and a deceleration of approximately 11g on reentry.

		(b) Qualify the posigrade rockets. (c) Qualify the recovery system. (d) Qualify the launch, tracking, and recovery phases of operation. (e) Qualify the Automatic Stabilization and Control System, including the Reaction Control System.
MR-2 Primate aboard	January 31, 1961	(a) Obtain physiological and performance data on a primate in ballistic space flight. (b) Qualify the Environmental Control System and aeromedical instrumentation. (c) Qualify the landing bag system. (d) Partially qualify the voice communication system. (e) Qualify the mechanically-actuated side hatch. (f) Obtain a closed-loop evaluation of the booster automatic abort system.
MR-BD Booster Development Flight	March 24, 1961	(a) Investigate corrections to booster problems as a result of the MR-2 flight. These problems were as follows: (1) Structural feedback to control system producing vane "chatter". (2) Instrument compartment vibration. (3) Thrust control malfunction.
MR-3 Manned	May 5, 1961	(a) Familiarize man with a brief but complete space flight experience including the lift-off, powered flight, weightless flight (for a period of approximately 5 minutes), reentry, and landing phases of the flight. (b) Evaluate man's ability to perform as a functional unit during space flight by: (1) Demonstrating manual control of spacecraft attitude before, during, and after retrofire. (2) Use of voice communications during flight. (c) Study man's physiological reactions during space flight. (d) Recover the astronaut and spacecraft.
MR-4 Manned	June 21, 1961	(a) Familiarize man with a brief but complete space flight experience including the lift-off, powered flight, weightless (for a period of approximately 5 minutes), atmospheric reentry and landing phases of the flight. (b) Evaluate man's ability to perform as a functional unit during space flight by: (1) Demonstrating manual control of spacecraft during weightless periods. (2) Using the

		spacecraft window and periscope for attitude reference and recognition of ground check points. (c) Study man's physiological reactions during space flights. (d) Qualify the explosively-actuated side egress hatch.
MERCURY-ATLAS MISSIONS AND TEST OBJECTIVES		
Mission	**Launch Date**	**Objectives**
Big Joe I	September 9, 1959	(a) To recover the spacecraft. (b) To determine the performance of the ablation shield and measure afterbody heating. (c) To determine the flight dynamic characteristics of the spacecraft during reentry. (d) To establish the adequacy of the spacecraft recovery system and procedures. (e) To establish the adequacy of recovery aids in assisting the recovery of the spacecraft. (f) To conduct familiarization of NASA operating personnel with Atlas launch procedures. (g) To evaluate the loads on the spacecraft during the actual flight environment. (h) To evaluate operation of the spacecraft control system.
MA-2	February 21, 1961	(a) To determine the integrity of the spacecraft structure, ablation shield, and afterbody shingles for a reentry from a critical abort. (b) To evaluate the performance of the operating spacecraft systems during the entire flight. (c) To determine the spacecraft full-scale motions and afterbody heating rates during reentry from a critical abort. (d) To evaluate the compatibility of the spacecraft escape systems with the Mercury-Atlas system. (e) To establish the adequacy of the location and recovery procedures. (f) To determine the closed-loop performance of the Abort Sensing and Implementation System. (g) To determine the ability of the Atlas booster to release the Mercury spacecraft at the position, altitude, and velocity defined by the guidance equations. (h) To evaluate the aerodynamic loading vibrational characteristics and structural integrity of

		the LO$_2$ boiloff valve, tank dome, spacecraft adapter and associated structures.
MA-4	September 13, 1961	(a) To demonstrate the integrity of the spacecraft structure, ablation shield, and afterbody shingles for a normal reentry from orbital conditions. (b) To evaluate the performance of the Mercury spacecraft systems for the entire flight. (c) To determine the spacecraft motions during a normal reentry from orbital conditions. (d) To determine the Mercury spacecraft vibration environment during flight. (e) To demonstrate the compatibility of the Mercury spacecraft escape system with the Mercury-Atlas system. (f) To determine the ability of the Atlas booster to release the Mercury spacecraft at the prescribed orbital insertion conditions. (g) To demonstrate the proper operation of the network ground command control equipment. (h) To evaluate the performance of the network equipment and the operational procedures used in establishing the launch trajectory and booster cutoff conditions and in predicting landing points. (i) To evaluate the applicable ground communications network and procedures. (j) To evaluate the performance of the network acquisition aids, the radar tracking system, and the associated operational procedures. (k) To evaluate the telemetry receiving system performance and the telemetry displays. (l) To evaluate the spacecraft recovery operations, as to the equipment and procedures used for communications and for locating and recovering the spacecraft, for a landing in the Atlantic Ocean along the Mercury network. (m) To obtain data on the repeatability of the booster performance of all Atlas missile and ground systems. (n) To determine the magnitude of the booster sustainer/vernier residual thrust after cutoff. (o) To evaluate the performance of the Abort Sensing and Implementation System. (p) To evaluate and develop applicable Mercury Network countdown and operational procedures. (q) To evaluate the Atlas booster with regard to engine start and potential causes for combustion

		instability.
MA-5 Primate aboard	November 29, 1961	(a) To demonstrate the performance of the Environmental Control System by utilizing a primate during an orbital mission. (b) To demonstrate satisfactory performance of the spacecraft systems throughout a Mercury orbital mission. (c) To determine by detail measurements, the heating rate and the thermal effects throughout the Mercury spacecraft for all phases of an orbital mission. (d) To exercise the satellite clock. (e) To determine the ability of the Atlas booster to release the Mercury spacecraft at the prescribed orbital insertion condition. (f) To demonstrate satisfactory performance of the Mercury Network in support of an orbital mission. (g) To demonstrate the ability of the Flight Controllers to satisfactorily monitor and control an orbital mission. (h) To demonstrate the adequacy of the recovery plans for an orbital mission; particular emphasis is required for the spacecraft occupant. (i) To evaluate the performance of the Abort Sensing and Implementation System. (j) To determine the magnitude of the booster sustainer/vernier residual thrust or impulse after cutoff. (k) To obtain data on the repeatability of the booster performance of all Atlas mission and ground systems. (l) To evaluate the Mercury Network countdown and operational procedures. (m) To evaluate the Atlas booster with regard to engine start and potential causes for combustion instability.
MA-6 Manned	February 20, 1962	(a) To evaluate the performance of man-spacecraft system in a three-orbit mission. (b) To evaluate the effects of space flight on the astronaut. (c) To obtain the astronaut's evaluation of the operational suitability of the spacecraft and supporting systems for manned space flight.
MA-7 Manned	May 24, 1962	(a) To evaluate the performance of man-spacecraft system in a three-pass orbital mission.

		(b) To evaluate the effects of space flight on the astronaut. (c) To obtain the astronaut's opinions on the operational suitability of the spacecraft systems. (d) To evaluate the performance of spacecraft systems replaced or modified as a result of previous missions. (e) To exercise and evaluate further the performance of the Mercury Worldwide Network.
MA-8 Manned	October 3, 1962	(a) To evaluate the performance of the man-spacecraft system in a six-pass orbital mission. (b) To evaluate the effects of an extended orbital space flight on the astronaut and to compare this analysis with those of previous missions and astronaut-simulator programs. (c) To obtain additional astronaut evaluation of the operational suitability of the spacecraft and support systems for manned orbital flights. (d) To evaluate the performance of spacecraft systems replaced or modified as a result of previous three-pass orbital missions. (e) To evaluate the performance of and exercise further the Mercury Worldwide Network and mission support forces and to establish their suitability for extended manned orbital flight.
MA-9 Manned	May 15, 1963	(a) To evaluate the effects on the astronaut of approximately one day in orbital flight. (b) To verify that man can function for an extended period in space as a primary operating system fo the spacecraft. (c) To evaluate in a manned one-day mission the combined performance of the astronaut and a Mercury spacecraft specifically modified for the mission.

Appendix 3 - Project Mercury Flight Data Summary

Flight	Launch date	Maximum altitude			Maximum range		Maximum velocity			Flight duration hr:min:sec
		Feet	Statute miles	Nautical miles	Statute miles	Nautical miles	Ft/sec earth-fixed	Ft/sec space-fixed	Mph space-fixed	
Big Joe 1	9-9-59	501,600	95.00	82.55	1,496.00	1,300.00	20,442	21,790	14,856.8	13:00
LJ-6	10-4-59	196,000	37.12	32.26	79.40	69.00	3,600	4,510	3,075.0	5:10
LJ-1A	11-4-59	47,520	9.00	7.82	11.50	10.00	2,040	2,965	2,021.6	8:11
LJ-2	12-4-59	280,000	53.03	46.08	194.40	196.00	5,720	6,550	4,465.9	11:06
LJ-1B	1-21-60	49,104	9.30	8.08	11.70	10.20	2,040	2,965	2,021.6	8:35
Beach abort	5-9-60	2,465	0.47	0.41	0.60	0.50	475	1,431	976.2	1:16
MA-1	7-29-60	42,768	8.10	7.04	5.59	4.85	1,560	2,495	1,701.1	3:18
LJ-5	11-8-60	53,328	10.10	8.78	13.60	11.80	1,690	2,618	1,785.0	2:22
MR-1A	12-19-60	690,000	130.68	113.56	234.80	204.00	6,350	7,200	4,909.1	15:45
MR-2	1-31-61	828,960	157.00	136.43	418.00	363.00	7,540	8,590	5,856.8	16:39
MA-2	2-21-61	602,140	114.04	99.10	1,431.60	1,244.00	18,100	19,400	13,227.3	17:56
LJ-5A	3-18-61	40,800	7.73	6.71	19.80	17.20	1,680	2,615	1,783.0	23:48
MR-BD	3-24-61	599,280	113.50	98.63	307.40	267.10	6,560	7,514	5,123.2	8:23
MA-3	4-25-61	23,760	4.50	3.91	0.29	0.25	1,135	1,726	1,176.8	7:19
LJ-5B	4-28-61	14,600	2.77	2.40	9.00	7.80	1,675	2,611	1,780.2	5:25
MR-3	5-5-61	615,120	116.50	101.24	302.80	263.10	6,550	7,530	5,134.1	15:22
MR-4	7-21-61	624,400	118.26	102.76	302.10	262.50	6,618	7,580	5,168.2	15:37
MA-4	9-13-61	750,300	142.10	123.49	26,047.00	22,630.00	24,389	25,705	17,526.0	1:49:20
MA-5	11-29-61	778,272	147.40	128.09	50,892.00	44,104.00	24,393	25,710	17,529.6	3:20:59
MA-6	2-20-62	856,279	162.17	140.92	75,679.00	65,763.00	24,415	25,732	17,544.1	4:55:23
MA-7	5-24-62	880,792	166.82	144.96	76,021.00	66,061.00	24,422	25,738	17,548.6	4:56:05
MA-8	10-3-62	928,429	175.84	152.80	143,983.00	125,118.00	24,435	25,751	17,557.5	9:13:11
MA-9	5-15-63	876,174	165.9	144.2	546,167.00	474,607.00	24,419	25,735	17,546.6	34:19:49

Listed range is earth track.
Big Joe = MA Development Flight.
MR-BD = Booster Development Flight.

LJ = Little Joe.
MR = Mercury Redstone.
MA = Mercury Atlas.
Flight duration is lift-off to impact.

Appendix 4 - Launch Site Summary
Cape Canaveral And Wallops Island

Space-craft No.	Use	Mission	Delivered to launch site	Date launched	Weeks of prep.	Remarks
1	Beach abort	System qualification test Escape landing Postlanding - unmanned	April 1, 1960	May 9, 1960	5	No launch vehicle used
2	MR-1A	Ballistic - unmanned	July 23, 1960	Dec. 19, 1960	21	Spacecraft reworked after MR-1 launch attempt where launch vehicle malfunctioned
3	LJ-5	Mercury max. dynamic pressure abort - unmanned	Sept. 27, 1960	Nov. 8, 1960	6	All objectives not accomplished
4	MA-1	Max. acceleration, max. heat on afterbody - unmanned	May 23, 1960	July 29, 1960	10	All systems not complete; all objectives not accomplished
5	MR-2	Ballistic - primate	Oct. 11, 1960	Jan. 31, 1961	16	Successful mission, Ham occupant
6	MA-2	Max. acceleration, max. heat on afterbody - unmanned	Sept. 1, 1960	Feb. 21, 1961	25	Successful mission
7	MR-3	Ballistic - manned	Dec. 9, 1960	May 5, 1961	21	First manned ballistic mission; mission successful; Shepard, pilot
8	MA-3	Orbital - unmanned	Nov. 18, 1960	April 25, 1961	23	Launched and aborted; all mission

						objectives not accomplished
8A	MA-4	Orbital - unmanned	May 11, 1961	Sept. 13, 1961	18	Spacecraft refitted and flown same configuration; mission successful
9	MA-5	Orbital - primate	Feb. 24, 1961	Nov. 29, 1961	40	Mission changed after delivery; 2 orbits flown successfully; Enos occupant
10	Ground test	Orbital flight Environmental test	Mar. 31, 1961	*June 1, 1962		Used at St. Louis in project orbit
11	MR-4	Ballistic - manned	Mar. 7, 1961	July 21, 1961	19	Manned ballistic mission; flown successfully; spacecraft lost on recovery; Grissom, pilot
12B	Unassigned	Orbital - manned 1-day mission				Mission canceled, spacecraft not delivered
13	MA-6	Orbital - manned	Aug. 27, 1961	Feb. 20, 1962	25	Unusually long pad period, 4 holds; first manned orbital mission; mission successful; Glenn, pilot
14	LJ-5A	Mercury max. dynamic pressure abort - unmanned	Jan. 20, 1961	Mar. 18, 1961	8	All mission objectives not accomplished
14A	LJ-5B	Mercury max. dynamic pressure abort - unmanned	April 4, 1961	April 28, 1961	3.5	Spacecraft refitted and flown same configuration; mission successful
15B	MA-10	Orbital -	Nov. 16,			Mission canceled

		manned 1-day mission	1962			after success of MA-9, spacecraft 20
16	MA-8	Orbital - manned	Jan. 16, 1962	Oct. 3, 1962	37	Mission changed from 3 to 6 orbital manned; mission successful; Schirra, pilot
17	Unas-signed	Orbital - manned 1-day mission	April 18, 1963			Delivered to Cape for parts support of manned 1-day missions
18	MA-7	Orbital - manned	Nov. 15, 1961	May 24, 1962	27	4 holds in pad period, 2nd 3 orbital mission; successful; Carpenter, pilot
19	Unas-signed	Orbital - manned	Mar. 20, 1962			Mission canceled after MA-8
20	MA-9	Orbital - manned 1-day mission	Oct. 9, 1962	May 15, 1963	31	22 orbital manned mission; mission successful; Cooper, pilot

* Completed ground test.

Appendix 5 - Project Mercury Budget Summary

SCHEDULE I

[Total Program, Obligations, and Expenditure by Fiscal Year as of January 31, 1963]

Fiscal Year 1959			
Function	Program	Obligation	Expenditure
Tracking and data acquisition, integrated system, study and test	0	0	0
Tracking and data acquisition, network operations	0	0	0
Spacecraft	16,768,469.98	16,768,650.95	16,768,650.95
Scout	3,932,000.00	3,932,000.00	3,932,000.00
Atlas	10,465,465.08	10,436,256.05	10,435,404.25
Little Joe	2,546,307.43	2,546,307.43	2,546,307.43
Redstone	8,967,182.00	8,967,182.00	8,967,182.00
Jupiter	1,779,153.97	1,779,153.97	1,779,153.97
Big Joe	437,211.87	437,211.87	437,211.87
Spacecraft support tech.	1,425,736.29	1,425,736.29	1,425,421.89
Flight operations	0	0	0
Recovery operations	0	0	0
Network operations	0	0	0
Network implementation	0	0	0
General administrative expense	34,462.59	34,462.59	34,462.59
Program overhead	115,374.75	115,374.75	115,374.75
Fiscal Year 1960			
Function	Program	Obligation	Expenditure
Tracking and data acquisition, integrated system, study and test	0	0	0
Tracking and data acquisition, network operations	65,784.75	65,784.75	65,784.75
Spacecraft	64,167,647.42	64,167,647.42	64,167,647.42
Scout	0	0	0
Atlas	11,496,686.97	11,496,686.97	11,496,680.97
Little Joe	10,865.38	10,865.38	10,865.38
Redstone	2,567,000.00	2,567,000.00	1,983,163.47
Jupiter	0	0	0

Big Joe	14,427.06	14,427.06	14,427.06
Spacecraft support tech.	2,490,090.81	2,488,421.01	2,485,117.33
Flight operations	0	0	0
Recovery operations	0	0	0
Network operations	0	0	0
Network implementation	0	0	0
General administrative expense	90,711.68	90,711.68	96,701.34
Program overhead	303,686.91	303,686.91	296,620.72
Fiscal Year 1961			
Function	**Program**	**Obligation**	**Expenditure**
Tracking and data acquisition, integrated system, study and test	400,000.00	400,000.00	400,000.00
Tracking and data acquisition, network operations	28,980.59	28,980.59	28,980.59
Spacecraft	55,718,834.77	54,755,159.29	43,359,310.50
Scout	0	0	0
Atlas	24,586,996.72	24,586,996.72	23,967,037.03
Little Joe	29,974.61	29,974.61	29,974.61
Redstone	2,241,553.22	2,239,553.22	2,092,583.89
Jupiter	0	0	0
Big Joe	0	0	0
Spacecraft support tech.	2,223,312.41	2,033,312.41	2,033,273.75
Flight operations	0	0	0
Recovery operations	2,621,496.73	2,621,496.73	2,429,675.70
Network operations	8,625,721.12	8,625,721.12	8,601,828.69
Network implementation	12,769,198.30	12,769,198.30	12,534,358.04
General administrative expense	355,825.71	355,825.71	428,247.47
Program overhead	1,191,242.61	1,191,242.61	1,095,561.44
Fiscal Year 1962			
Function	**Program**	**Obligation**	**Expenditure**
Tracking and data acquisition, integrated system, study and test	0	0	0
Tracking and data acquisition, network operations	0	0	0
Spacecraft	1,435,338.89	1,238,708.29	831,673.28
Scout	0	0	0
Atlas	5,540,000.00	5,540,000.00	5,540,000.00
Little Joe	0	0	0
Redstone	0	0	0

Jupiter	0	0	0
Big Joe	0	0	0
Spacecraft support tech.	890,659.25	890,659.25	335,216.58
Flight operations	1,316,784.83	1,245,376.32	1,141,132.23
Recovery operations	6,312,134.32	5,851,827.00	5,832,311.71
Network operations	0	0	0
Network implementation	0	0	0
General administrative expense*	5,874,793.40	5,814,037.00	5,450,211.30
Program overhead*	2,429,177.62	2,232,327.91	1,872,758.60

Fiscal Year 1963			
Function	Program	Obligation	Expenditure
Spacecraft	185,000.00	18,384.05	1.05
Support development	75,000.00	71,323.42	29,174.09
Operations:	3,082,000.00		
Flight		532,395.27	77,384.85
Recovery		2,531,421.92	269,456.64

*Computed using a percentage of 86 which was the percent of program overhead applied directly to Project Mercury as of June 30, 1962.

SCHEDULE II

[Total Program, Obligations, and Expenditure by Fiscal Year as of January 31, 1963]

Fiscal Year 1959			
Function	Program	Obligation	Expenditure
Salaries and expenses	1,332,170.95	1,332,170.95	1,332,170.95

Fiscal Year 1960			
Function	Program	Obligation	Expenditure
Salaries and expenses	5,000,648.87	5,000,648.87	5,000,648.87

Fiscal Year 1961			
Function	Program	Obligation	Expenditure
Salaries and expenses	7,747,483.15	7,747,703.30	7,747,703.30

Fiscal Year 1962			
Function	Program	Obligation	Expenditure
Salaries and expenses*	1,332,170.95	1,332,170.95	1,332,170.95

*Based on a percentage of 62 which was the percent of program overhead applied to Project Mercury from this appropriation as of June 30, 1962.

SCHEDULE III

[Total Program, Obligations, and Expenditure by Fiscal Year as of January 31, 1963]

Fiscal Year 1960			
Function	Program	Obligation	Expenditure
Equipment and instrumentation	500,000.00	499,700.00	499,700.00

Appendix 6 - Location of Mercury Spacecraft and Exhibit Schedule
[July 1963]

Spacecraft No. 1 - beach abort and sled test (present status - display at MSC Public Affairs Office)		
	August 9-18, 1963 -	Springfield, Illinois Illinois State Fair
	September 9-14, 1963 -	Abilene, Texas West Texas State Fair
	October 3-4, 1963 -	Memphis, Tennessee Education Workshop
	November 1963 -	Available for exhibit scheduling or release to the U.S. Air Force Museum, Washington, D.C.
Spacecraft No. 2 - MR-1 ballistic, unmanned and RCS test bed (present status - display at MSC Public Affairs Office)		
	August 1-5, 1963 -	Kansas City, Missouri U.S. Air Force Industrial Propulsion Richards-Gebaur Air Force Base, Missouri
	August 17-18, 1963 -	Kalamazoo, Michigan Air Show
	August 20-24, 1963 -	Champaign, Illinois National Aero Scout Conference University of Illinois
	August 31-September 3, 1963 -	Dallas, Texas Dallas Civil Air Patrol
	September 16-21, 1963 -	Texarkana, Texas Four State Fair and Rodeo
	October 5-10, 1963 -	Houston, Texas International Association of Chiefs of Police
	October 15, 1963 -	Available for exhibit scheduling or release to Goddard Museum, New Mexico
Spacecraft No. 3 - Little Joe 5 (lost on launch)		
Spacecraft No. 4 - MA-1 ballistic, unmanned (lost on launch)		
Spacecraft No. 5 - MR-2 ballistic, primate and net couch qualification (present status - MSC-Hanger 135 - being refurbished after couch program)		
	September 1963 -	Release to Marshall Space Flight Center, Huntsville, Alabama, for display
Spacecraft No. 6 - MA-2 ballistic, unmanned (present status - permanent item at		

Houston, Texas, World Trade Center)		
Spacecraft No. 7 - MR-3 ballistic, manned (Shepard) (present status - display at Smithsonian Institution)		
Spacecraft No. 8 - MA-3, MA-4 orbital, unmanned (present status - storage at MSC)		
	October 5-20, 1963 -	Dallas, Texas Texas State Fair
	October 28-30, 1963 -	Chicago, Illinois National Electronics Conference
	November 1963-July 1964 -	Houston and Texas special school tour
	September 10-14, 1964 -	Abilene, Texas West Texas State Fair
	September 17-22, 1964 -	Waco, Texas Heart of Texas Fair
Spacecraft No. 9 - MA-5 orbital, primate and project orbit (present status - testing at MSC); available for display upon completion of testing		
Spacecraft No. 10 - Project orbit and R&D test bed (not flown) (present status - testing at MSC); available for display upon completion of testing		
Spacecraft No. 11 - MR-4 ballistic, manned (Grissom) (lost on recovery)		
Spacecraft No. 12 - MA-6 backup and manned 1-day mission (present status - storage at MSC); available for exhibit scheduling by NASA Headquarters		
Spacecraft No. 13 - MA-6 orbital, manned (Glenn) (present status - display at Smithsonian Institute)		
Spacecraft No. 14 - Little Joe 5B maximum dynamic pressure abort and parachute qualification tests (present status - storage at MSC); release to Langley Research Center		
Spacecraft No. 15 - Manned 1-day mission (not flown) (present status - storage at Cape Canaveral); available for exhibit scheduling by NASA Headquarters		
Spacecraft No. 16 - MA-8 orbital, manned (Schirra) (present status - United States Information Agency tour of France)		
	August- September 1963 -	Athens, Greece World Scout Jamboree Foreign Tour
	October 1963 -	Orlando, Florida Space Science Achievements
	November 1963 -	Amsterdam, Holland Amsterdam Exposition
	December 1963-April	World Tour

	1964 -	
	May 1964 -	Boston, Massachusetts Peaceful Uses of Space Conference

Spacecraft No. 17 - Manned 1-day mission (not flown) (present status - storage at MSC); available for exhibit scheduling by NASA Headquarters

Spacecraft No. 18 - MA-7 orbital, manned (Carpenter) (present status - display at NASA Headquarters)

	August 10-14, 1963 -	Houston, Texas American Municipal Association
	August 21-24, 1963 -	Boulder, Colorado
	September 20-29, 1963 -	Houston, Texas Houston International Trade and Travel Fair
	October 23-November 2, 1963 -	Jacksonville, Florida Agriculture and Industrial Fair
	November 21-23, 1963 -	Baton Rouge, Louisiana Aerospace Education Workshop
	December 1963-April 1964 -	Available for exhibit scheduling by NASA Headquarters
	May 1964 -	International Science Fair (location unknown at this time)
	June 1964- June 1965 -	United States Tour
	July 1965 -	Release to Cape Canaveral

Spacecraft No. 19 - MA-8 backup (not flown) (present status - storage at MSC); available for display

Spacecraft No. 20 - MA-9 orbital, manned (Cooper) (present status - display at NASA Headquarters)

	September 21-29, 1963 -	Oklahoma City, Oklahoma Oklahoma State Fair
	September 30-October 2, 1963 -	Houston, Texas Mercury Report Conference
	October 3, 1963 -	Available for 50-State capitol tour

Big Joe Boilerplate - Big Joe ballistic unmanned (present status - display at MSC-PAO)

	October 1-5, 1963 -	Waco, Texas Heart of Texas Fair
	October 10, 1963-June	Retained by MSC for Houston and Texas exhibits

1965 -

SUGGESTED FINAL DISPOSITION OF MANNED MERCURY SPACECRAFT

Spacecraft No. 16 - MA-8 - orbital, manned (Schirra); after July 1965 - Undetermined

Spacecraft No. 18 - MA-7 - orbital, manned (Carpenter); after July 1965 - Cape Canaveral

Spacecraft No. 20 - MA-9 - orbital, manned (Cooper); after July 1965 - Houston, Texas, for permanent installation at MSC

SUGGESTED FINAL DISPOSITION OF UNMANNED MERCURY SPACECRAFT

Spacecraft No. 1 or **No. 2** - after January 1964 - Goddard Museum in New Mexico

Spacecraft No. 1 or **No. 2** - after January 1964 - U.S. Air Force Museum, Washington, D.C.

Spacecraft No. 5 - MR-2 - ballistic, primate and net couch qualification - immediate release to Marshall Space Flight Center

Spacecraft No. 14 - Little Joe 5B maximum dynamic pressure and parachute qualification tests - immediate release to Langley Research Center

Appendix 7 - Launch Vehicle Deliveries To Cape Canaveral

Launch vehicle	Delivery date	Mission	Comments
Redstone			
1	August 3, 1960	MR-1	Failed on pad
2	December 20, 1960	MR-2	
3	December 3, 1960	MR-1A	Replaced launch vehicle No. 1
5	March 7, 1961	MRBD	Booster Development Flight
7	March 30, 1961	MR-3	
8	June 12, 1961	MR-4	
4			Not used
6			Not used
Atlas			
50-D	June 18, 1960	MA-1	
67-D	September 20, 1960	MA-2	
100-D	March 14, 1961	MA-3	
88-D	July 15, 1961	MA-4	
93-D	October 9, 1961	MA-5	
109-D	November 30, 1961	MA-6	
107-D	March 6, 1962	MA-7	
113-D	August 8, 1962	MA-8	
20-D			Assigned to NASA Hq.
10-D		Big Joe	
130-D	March 18, 1963	MA-9	
144-D		MA-10	Canceled
152-D			In storage - unassigned
77-D			Canceled
103-D			Canceled
167-D			Canceled

Appendix 8 - Key Management Progression Involving Project Mercury

January 26, 1959: *Space Task Group*
    Project Manager - Robert R. Gilruth
        Assistant Project Manager - Charles J. Donlan
        Special Assistant to Project Manager - Paul E. Purser
        Technical Services Assistant to Project Manager - Jack A. Kinzler
        Administrative Assistant and Head of Staff Services - Paul D. Taylor
        Chief of Flight Systems Division - Maxime A. Faget
        Chief of Operations Division - Charles W. Mathews
        Chief of Engineering and Contract Administration Division - Charles H. Zimmerman
        Services Representatives - Lt. Colonel Keith Lindell, U.S. Air Force; Lt. Colonel Martin Raines, U.S. Army; and Commander Paul Havenstein, U.S. Navy
        Aero Medical Consultants - Dr. Stanley White, U.S. Air Force; Dr. William Augerson, U.S. Army; and Dr. Robert Voas, U.S. Navy

February 16, 1959:
    McDonnell Aircraft Corporation Liaison Officer and Test Program Coordinator to STG - Frank G. Morgan

June 19, 1959:
    Capsule Coordination Office:
        Group 1 - Loads, thermodynamics, structures, and aerodynamics
        Group 2 - Cabin, life support, and controls
        Group 3 - Electronics, recovery, and sequencing
        Group 4 - Mechanical handling and transport equipment; schedules; hardware testing, spares, and standards and specifications

August 3, 1959: *Space Task Group*
    Director of Project Mercury - Robert R. Gilruth
        Assistant Director of Project Mercury - Charles J. Donlan
        Staff Assistants:
            Special Assistant - Paul E. Purser
            Technical Assistant - James A. Chamberlin
            Executive Assistant - Raymond L. Zavasky

U.S. Air Force - Colonel Keith Lindell
U.S. Army - Lt. Colonel Martin Raines
U.S. Navy - Commander Paul Havenstein
Langley Research Center - Kemble Johnson
Public Affairs Officer - Lt. Colonel John A. Powers
Staff Services:
    Personnel Assistant - Burney H. Goodwin
    Administrative Services - Guy W. Boswick, Jr.
Technical Services - Jack A. Kinzler
Astronaut and Training Group - Colonel Keith Lindell
Flight Systems Division - Maxime A. Faget
    Computing Group - Katherine S. Stokes
    Systems Test Branch - William M. Bland, Jr.
    Performance Branch - Aleck C. Bond
        Aerodynamics Section - Alan B. Kehlet
        Loads Section - George A. Watts
        Heat Transfer Section - Leonard Rabb
    Dynamics Branch - Robert G. Chilton
        Flight Control Section - Richard R. Carley
        Space Mechanics Section - Robert G. Chilton (Acting)
    Life Systems Branch - Dr. Stanley C. White
    On-Board Systems Branch - H. H. Ricker, Jr.
        Electrical Systems Section - H. H. Ricker, Jr.
        Mechanical Systems Section - J. B. Lee
Operations Division - Charles W. Mathews
    Assistant Chief for Implementation - G. Merritt Preston
    Assistant Chief for Plans and Arrangements - Christopher C. Kraft, Jr.
    Assistant to Division Chief - John D. Hodge
    Mission Analysis Branch - John P. Mayer
        Trajectory Analysis Section - John P. Mayer (Acting)
        Operational Analysis Section - Jack Coehn (Acting)
        Mathematical Analysis Section - Stanley H. Cohn
    Flight Control Branch - Gerald W. Brewer
        Control Central and Flight Safety Section - Gerald W. Brewer (Acting)
        Training Aids Section - Harold I. Johnson
    Launch Operations Branch - Charles W. Mathews (Acting)
        AMR Project Office - Elmer H. Buller
        Preflight Checkout Section - G. Merritt Preston (Acting)
    Recovery Operations Branch - Robert F. Thompson
    Engineering and Contract Administration Division - James A. Chamberlin (Acting)
        Field Representative at McDonnell - Wilbur H. Gray
        Capsule Coordination Office - James A. Chamberlin
        Contracts and Scheduling Branch - George F. McDougall, Jr.
        Engineering Branch - Caldwell Johnson

August 10, 1959 (Changes):
    Operations Division:
        AMR Project Office - deleted from Launch Operations Branch and placed directly under division head
        Preflight Checkout Section - deleted from Launch Operations Branch

September 15, 1959:
    Office of the Director:
        Associate Director (Operations) - established. Walter C. Williams appointed to position. Charles J. Donlan became Associate Director (Development)

November 16, 1959 (Changes):
    Engineering and Contract Administration Division:
        Contracts and Scheduling Branch:
            Contracts Section - Joseph V. Piland (Acting)
            Scheduling Section - Nicholas Jevas

November 23, 1959 (Changes):
    Staff Services - abolished
    Administrative Services Office - established under Director - no change in personnel
    Personnel Office - established under Director - no change in personnel
    Procurement and Supply Office - Glenn F. Bailey
    Budget and Finance Office - John P. Donovan

January 11, 1960 (Changes):
    Engineering and Contract Administration Division - redesignated as Engineering Division - James A. Chamberlin, Acting Chief, was appointed as Chief and relieved of duties as Technical Assistant to Director
    Technical Assistant - Kenneth S. Kleinknecht appointed to Director's staff
    Staff Assistant - Martin C. Byrnes, Jr. appointed to Director's staff

January 18, 1960:
  Mercury-Atlas Flight Test Working Group - B. Porter Brown (Chairman)

February 12, 1960 (Operational Organization and Appointments):
  Operations Director - Walter C. Williams
    Flight Director - Christopher C. Kraft, Jr.
    Chief Flight Surgeon - Dr. Stanley C. White
    Launch Operations Manager - G. Merritt Preston
    Capsule Operations Manager - Scott H. Simpkinson

  (Note: These appointments did not affect the status of individuals in the formal organization)

February 16, 1960:
  Digital Computer Group - established under Director - Stanley H. Cohn appointed Head and relieved of duties as Head of Mathematical Analysis Section
  Operations Division - Mission Analysis Branch:
    Mathematical Analysis Section - John P. Mayer (Acting)

May 12, 1960:
  NASA Resident Representative at McDonnell Aircraft - position was formally organized under the Engineering and Contract Administration Division and designated Field Representative at McDonnell. The position was now placed under the Director, and there was no change in personnel
    Resident Systems Test Engineer - T. M. Edwards
    Resident Inspection Engineer - J. C. Moser
    Inspectors - subsequently assigned

May 25, 1960:
  Advanced Vehicle Team - Robert O. Piland

  (Note: Individuals appointed were charged with preliminary design studies leading to establishment of requirements for an advanced multi-man space vehicle. Team membership did not affect status of individuals in formal organization)

September 1, 1960:
    Security Office - Donald D. Blume - activity established under Director
    Flight Systems Division:
        Systems Test Branch - redesignated Project Engineering Branch and transferred to the Engineering Division
    Engineering Division:
        Project Engineering Branch - William M. Bland, Jr., also appointed Assistant Chief of Division
    Flight Systems Division (Reorganization - changes only):
        Assistant Chief for Mercury Support - Aleck C. Bond
        Apollo Projects Office - Robert O. Piland
        Systems Test Branch - deleted
        Electrical Systems Branch - formerly a section under On-Board Systems Branch:
            Communications Systems Section - Ralph S. Sawyer
            Instrumentation Systems Section - Alfred Eickmeier
        Flight Dynamics Branch - formerly Dynamics Branch:
            Flight Control Branch - Thomas V. Chambers
            Dynamics Analysis Section - Richard R. Carley
            Aerodynamics Section - Bruce G. Jackson

        (Note: Flight Control Section and Space Mechanics Section maintained status quo)
        Life Systems Branch - Dr. Stanley C. White
            Aerospace Medical Section - James P. Henry
            Crew Equipment Section - Richard S. Johnson (Acting)
        Systems Engineering Branch - Caldwell Johnson:
            Systems Integration Branch - Owen E. Maynard
            Equipment Engineering Section - Richard F. Smith
            Mechanical Systems Section - Richard B. Ferguson
        Structures Branch - Robert E. Vale:
            Structural Analysis Section - Robert E. Vale (Acting)
            Loads Section - George A. Watts
            Heat Transfer Section - Kenneth C. Weston
        On-Board Systems Branch - abolished

December 8, 1960:
    Office of NASA - Space Task Group Field Representative at McDonnell Aircraft Corporation - (Reorganization - no change in personnel)
        Capsule Systems Test Coordinator - Thomas M. Edwards
        Capsule Systems Test Engineers - Phillip M. Deans, Archibald E. Morse, and Louis Leopold
        Coordination - Harle L. Vogel
        Inspectors - Albert J. Eaton (lead)
        Test and Schedules - William J. Nesbitt
        Consultant - Scott H. Simpkinson

December 9, 1960:
    Flight Systems Division:
        Life Systems Branch - Redesignated Life Systems Group and transferred to Office of the Director. The Aerospace Medical Section and Crew Equipment Section remained under the Group, and there was no change in personnel

January 16, 1961:
    Flight Systems Division - Flight Dynamics Branch:
        Dynamic Analysis Section - Redesignated as the Navigation and Guidance Section
        Flight Control Section - Redesignated as the Attitude Control Section

March 21, 1961:
    Operations Division - Launch Operations Branch:
        Established as the Preflight Operations Division with G. Merritt Preston as Chief
        Flight Operations Division - Established from the remaining elements of the Operations Division, with the exception of the Mercury-Atlantic Missile Range Projects Office. Charles W. Mathews was appointed Division Chief
        Mercury-Atlantic Missile Range Projects Office - Designated a staff function of the Director's office. No change in personnel
        Life Systems Group - Accorded division status as the Life Systems Division. No change in personnel
        Reliability and Flight Safety Office - Established as a staff function of the Director's office. Frederick J. Bailey was appointed office head
        Business Management Office - Designated on organization chart with the following sub-offices: Procurement and Supply Office; Personnel Office; Budget and Finance Office; Administrative Services Office; and Security Office. No personnel assignment was indicated for the Chief of the Business Management Office

April 1, 1961:
    Office of the Director:
        Associate Director NASA - Space Task Group - Walter C. Williams. Charles J. Donlan joined the staff of the Langley Research Center in the position of Associate Director as of this date

April 13, 1961:
Business Management Office - Designation changed to Office of Assistant Director for Administration with Wesley L. Hjornevik as Assistant Director

June 5, 1961:
Flight System Division:
Flight Vehicles Integration Branch - Established with Maxime A. Faget, the Division Chief, serving as Acting Head

June 8, 1961:
Facilities Planning Panel - Established to identify and coordinate design requirements for the permanent facilities of the Manned Spacecraft Center. Raymond L. Zavasky was appointed Chairman

July 10, 1961:
Architect-Engineer Selection Board - Established to review the qualifications of an adequate number of architect-engineer firms when procurement by contract of architect-engineer services for a particular project was contemplated. I. E. Campagna was appointed Chairman

July 31, 1961:
Office of Assistant Director for Administration - Management Services Office - Established with Philip H. Whitbeck appointed as Chief
Personnel Office - Stuart H. Clark

August 24, 1961:
Manager of Space Task Group Cape Operations - G. Merritt Preston, with exception of periods when Walter C. Williams is present. Assignment was in addition to Preston's duties as Chief, Preflight Operations Division

September 6, 1961:
Office of Assistant Director for Administration - Management Services Office:
Management Analysis Office - Established with Charles F. Bingman appointed as Head (Acting)

September 19, 1961:
    Office of Assistant Director for Administration - Management Services Office:
        Transportation Office - Established with Edward Johnson designated as Chief

September 29, 1961:
    Office of Assistant Director for Administration - Management Services Office:
        Supply Office - Established with Thomas J. Porter designated as Chief (Acting)

October 2, 1961:
    Engineering Division:
        Engineering Data Office - Established with Robert E. McKann designated Office Head

November 1, 1961:
    Space Task Group - Redesignated Manned Spacecraft Center

January 15, 1962:
    Mercury Project Office - Established with Kenneth S. Kleinknecht appointed as Manager. Organization and staff were subsequently announced
    Apollo Spacecraft Project Office - Established with Charles W. Frick as Manager

January 15, 1962:
    Gemini Spacecraft Project Office - Established with James A. Chamberlin appointed as Manager

January 15, 1962:
    Flight Systems Division - Abolished
    Office of Assistant Director for Research and Development - Maxime A. Faget
        Spacecraft Research Division - Charles W. Mathews
        Life Systems Division - Dr. Stanley C. White
        Systems Evaluation and Development Division - Aleck C. Bond
        Space Physics Division - Not organized

January 31, 1962:
    Engineering Division - Abolished. Workload divided among Mercury, Gemini, and Apollo Project Offices

January 31, 1962:
    Mercury Project Office - Organized and manned:
    Office of Project Manager:
        Deputy Project Manager - William M. Bland, Jr.
        Project Engineering Office - Edison M. Fields
        Project Engineering Field Office (Cape) - A. E. Morse, Jr.
        Engineering Operations Office - Joseph V. Piland (Acting)
        Engineering Data and Measurements Office - Robert E. McKann

February 26, 1962:
    Office of Assistant Director for Operations - Flight Crew Operations Division - Established with Warren J. North appointed as Chief

May 21, 1962:
    Office of Assistant Director for Operations:
        Flight Operations Division - Reorganization. Christopher C. Kraft, Jr. remained Chief
        Technical Assistants - Sigurd A. Sjoberg and Robert D. Harrington
        Executive Assistant - Chris C. Critzos
        Assistant Chief for Flight Control - John D. Hodge
            Flight Control Operations Branch - John D. Hodge (Acting)
                Flight Operations Section - Eugene F. Kranz (Acting)
                Systems Analysis Section - Arnold D. Aldrich
                Training and Simulation Section - Robert E. Ernull
            Operational Facilities Branch - Howard C. Kyle
                RF Systems Section - James K. Meson
                Information Flow Section - Dennis E. Fielder
                Network Requirements Section - Thomas Stuart
            Mission Control Center Branch - Tecwyn Roberts
                Control Center Design Section - Richard A. Hoover
                Mission Logic and Computer Hardware Section - Glynn S. Lunney
                Simulation Design Section - Harold G. Miller
            Data Coordination Branch - Richard G. Arbic
        Assistant Chief for Mission Planning - John P. Mayer
            Real Time Program Development Branch - Lynwood C. Dunseith

            Gemini Program Development Section - La Rue W. Burbank
            Apollo Program Development Section - Unassigned
        Operations Analysis Branch - Carl R. Huss
            Prelaunch Mission Analysis Section - Charlie C. Allen
            Postflight Trajectory Analysis Section - Donald J. Incerto
            Performance and Guidance Section - Marlowe D. Cassetti
        Mathematical Physics Branch - Edward A. Knobelauch
            Mathematical Support Section - Paul G. Brumberg
            Computer Operation and Programing Section - I. Edna Hawkins
            Advanced Mission Analysis Section - Unassigned
        Mission Analysis Branch - Morris V. Jenkins
            Lunar Trajectory Section - Harold D. Beck
            Reentry Studies Section - John R. Gurley, Jr.
        Rendezvous Analysis Branch - James F. Dalby (Acting)
            Earth Rendezvous Section - John E. Gerstle, Jr.
            Lunar Rendezvous Section - Unassigned
    Assistant Chief for Operational Support - Robert F. Thompson
        Recovery Branch - Robert F. Thompson (Acting)
            Current Operations Section - John B. Graham, Jr.
            Advanced Planning Section - William C. Hayes, Jr.
        Operational Evaluation and Test Branch - Peter J. Armitage
            Systems Suitability Section - Peter J. Armitage (Acting)
            Test and Development Section - Milton L. Windler

June 15, 1962:
    Office of Reliability and Flight Safety - Frederick J. Bailey
        Design Evaluation Staff:
            Communications and Telemetry - Lawrence Steinhardt
            Electrical and Power Supply - Lawrence Steinhardt (Acting)
            Instrumentation - Lawrence Steinhardt (Acting)
            Navigation and Guidance - George S. Shigekawa (Acting)
            Pyrotechnics - George S. Shigekawa (Acting)
            Environmental Control System - John C. French (Acting)
            Propulsion - John W. Conlon (Acting)
            Human Factors - Thomas J. Edwards (Acting)
            Structure and Ablation Shield - John C. French (Acting)
            Landing Systems - John C. French (Acting)
            Separation Devices - George S. Shigekawa (Acting)
            Electrical Sequential Systems - Thomas J. Edwards (Acting)
            Analytical Methods - Thomas J. Edwards
            Quality Assurance - Karl P. Sperber
        Operations Evaluation Staff:

    Mechanical and Interface - Norbert B. Vaughn
    Electrical and Power Supply - Norbert B. Vaughn (Acting)
    Propulsion and Pyrotechnics - John W. Conlon
    Flight and Checkout Procedures - Frederick J. Bailey (Acting)
Project Reliability Evaluation Staff:
    Mercury Reliability Advisor - Charles Rice
    Gemini Reliability Advisor - Lemeul Menear
    Apollo Reliability Advisor - George S. Shigekawa

June 25, 1962:
Office of Assistant Director for Operations - Aerospace Medical Operations Office - Established with Dr. Charles A. Berry appointed as Chief

September 18, 1962:
Coordinator of Astronaut Affairs - Astronaut Donald K. Slayton

October 29, 1962:
Assistant Director for Information and Control Systems - G. Barry Graves

Appendix 9 - Contractors and Subcontractors Supporting Project Mercury

Research and development of Project Mercury was literally a national effort, especially when considering the number of contractors, subcontractors, and suppliers involved in the program. McDonnell, as a prime contractor for the spacecraft, was assisted by an estimated 4,000 suppliers and subcontractors throughout the country. Also of significant importance was the industrial effort put forth in the manufacture of launch vehicles and ground support equipment, and the construction of the worldwide tracking network in support of Project Mercury. Quite probably, if all were combined, the total figure would reach 10,000. Obviously, a complete listing is beyond the scope or intent of this work, but herein is a cross section of contributory effort covering the major systems and components.

McDonnell Aircraft Corporation, St. Louis, Missouri -
  Prime contractor for the Mercury spacecraft.
Convair Astronautics Division (GD/A), San Diego, California -
  Prime contractor for the Atlas launch vehicle system used for the manned orbital phase of Project Mercury. Procured through Space Systems Division of Air Force Systems Command.
Chrysler Corporation Missile Division, Detroit, Michigan -
  Prime contractor for the Redstone launch vehicle system used in the manned suborbital phase of Project Mercury. Procured through Army Missile Command.
North American Aviation, Inc., El Segundo, California -
  Contractor for the Little Joe launch vehicle airframe used in aerodynamic and abort technique testing program phase of Project Mercury.
Ventura Division (formerly Radioplane) of the Northrop Corporation, Van Nuys, California -
  Contractor for the Mercury spacecraft landing and recovery system.
B. F. Goodrich Company, Akron, Ohio -
  Contractor for the Mercury spacecraft astronaut pressure suit.
Western Electric Company, New York City, New York -
  Prime contractor for the Mercury worldwide tracking network.
Minneapolis-Honeywell Regulator Company, Minneapolis, Minnesota -
  Stabilization system for the Mercury spacecraft.
Bell Aerospace Corporation, Buffalo, New York -
  Reaction control system for the Mercury spacecraft.
AiResearch Manufacturing Division of the Garrett Corporation, Los Angeles, California -
  Environmental control system for the Mercury spacecraft.
The Perkin-Elmer Corporation, Norwalk, Connecticut -
  Periscope for the Mercury spacecraft.
Eagle-Picher Company, Joplin, Missouri -
  Batteries for the Mercury spacecraft.
Barnes Engineering Company, Stamford, Connecticut -
  Horizon scanner for the Mercury spacecraft.
Wheaton Engineering Corporation, Wheaton, Illinois -

Time-delay relays and programer for Mercury spacecraft.

Donner Scientific Company, Concord, California -
  Maximum altitude sensor and thrust cutoff sensor.

Atlantic Research Corporation, Alexandria, Virginia -
  Escape tower jettison rocket and posigrade rocket for the Mercury spacecraft.

Thiokol Chemical Corporation, Elkton, Maryland -
  Retrograde rocket for Mercury spacecraft.

Lockheed Propulsion Company, Redlands, California -
  Rocket motor for the Mercury spacecraft escape tower. (The original contract for this component was signed with the Grand Central Rocket Company, Redlands, California, which was later purchased by Lockheed.)

Cincinnati Testing and Research Laboratory of the Studebaker-Packard Corporation, Cincinnati, Ohio -
  Heat shield for the Mercury spacecraft.

Walter Kidde Company, Belleville, New Jersey -
  Emergency flotation bag for the Mercury spaceccraft.

Aeronca Corporation, Middletown, Ohio -
  Honeycomb panels for the Mercury spacecraft impact landing support.

Collins Radio Corporation, Cedar Rapids, Iowa -
  HF and UHF voice communications and UHF recovery antenna, onboard communications for the Mercury spacecraft.

Motorola, Incorporated, Franklin Park, Illinois -
  Command receivers, onboard communications for Mercury spacecraft.

Texas Instruments, Incorporated, Dallas, Texas -
  Onboard telemetry communications for Mercury spacecraft.

Cooper Electric Company -
  Minitrack beacon, onboard communications for Mercury spacecraft.

Melpar, Incorporated, Falls Church, Virginia -
  C- and S-band antennas, onboard communications for Mercury spacecraft.

Avion Division and G. E. -
  C- and S-band beacons, onboard communications for Mercury spacecraft.

Consolidated Electrodynamics Corporation, St. Louis, Missouri -
  Tape recorder for Mercury spacecraft.

Electro-Voice and R. E. Darling Companies, Bethesda, Maryland -
  Communication devices for Mercury astronaut pressure suit.

D. B. Milliken Company, Arcadia, California -
  Camera for Mercury spacecraft.

Waltham Precision Instrument Company, Waltham, Massachusetts -
  Satellite clock for Mercury spacecraft. (Contract was canceled on December 14, 1960, and the component was replaced with the orbital timing device fabricated by McDonnell Aircraft Corp.)

Bendix Radio Division of the Bendix Corporation, Baltimore, Maryland -
  Ground-air communications, radar, and aquisition systems for the Mercury worldwide tracking network.

Bendix-Pacific Division of the Bendix Corporation, North Hollywood, California -
  Telemetry, antennas, displays and radar data processing for the Mercury worldwide tracking network.
International Business Machines Corporation, New York, New York -
  Computers and computer programming for the Mercury worldwide tracking network.
Burns and Roe, Incorporated, New York, New York -
  Architecture, site engineering, and logistics for the Mercury worldwide tracking network.
Stromberg-Carlson (Division of General Dynamics), Rochester, New York -
  Control center consoles for the Mercury worldwide tracking network.
Lincoln Laboratory of the Massachusetts Institute of Technology, Lexington, Massachusetts -
  Technical consulting services to NASA on all phases of tracking and computing.
Space Electronics Corporation -
  Study contract leading to design of Mercury Control Center.
Aeronutronics Systems, Incorporated (Division of Ford Motor Company), Los Angeles, California -
  Study contract relative to tracking and computing problems.
Grumman Aircraft Engineering Corporation, Bethpage, New York -
  Operations analysis study of recovery problems associated with a three-orbit mission.
Tenney Engineering, Incorporated, Union, New Jersey -
  Environmental test chamber constructed in Hanger S, Cape Canaveral, Florida.
Philco Corporation, Philadelphia, Pennsylvania -
  Range monitors for Mercury worldwide tracking network.
Pan American Airways, Cape Canaveral, Florida -
  Atlantic Missile Range operations in support of Project Mercury.
Inter-Electronics Corporation, New York, New York -
  Static inverters for the Mercury spacecraft.
Amp, Incorporated, Greenwich, Connecticut -
  Design of stationary egress system.
Federal Electric Corporation, Cape Canaveral, Florida -
  Space telemetry.
Space Technology Laboratories, Redondo Beach, California -
  Analysis of flight instrumentation and design trajectories for Mercury-Atlas program.

Appendix 10 - Government Organizations Supporting Project Mercury

Headquarters, National Aeronautics and Space Administration, Washington, D.C. -
General manager of national space program.

Manned Spacecraft Center (formerly Space Task Group), Houston, Texas -
Project manager for manned space flight program.

Langley Research Center, Langley Field, Virginia -
Wind tunnel aerodynamics, fluid mechanics, stability and control, vibration and flutter, loads, structures, materials, arc-jet, reentry body, and planning and contracting for Mercury instrumentation facilities.

Launch Operations Division (Later Launch Operations Center), Cape Canaveral, Florida -
Launch vehicle manager for Mercury-Redstone.

Lewis Research Center, Cleveland, Ohio -
Flight propulsion, engine testing, stabilization and control system for spacecraft.

Ames Research Center, Moffett Field, California -
Basic and applied research on aeronautical and space problems, atmosphere reentry forms and high-speed aerodynamic research.

Marshall Space Flight Center (formerly Development Operations Division of ABMA), Huntsville, Alabama -
Spacecraft launch vehicle development and reliability, and trajectory studies for Mercury-Redstone.

Wallops Station, Wallops Island, Virginia -
Little Joe flight test for Mercury spacecraft and launch test of Mercury model spacecraft.

Goddard Space Flight Center, Greenbelt, Maryland -
Coordinated all tracking and processed all tracking data for Mercury-Redstone and Mercury-Atlas flights.

Flight Research Center Edwards, California -
High-speed flight research and drogue parachute tests.

Department of Defense, Washington, D.C. -
Launch and recovery operations.

Weather Bureau of the Department of Commerce, Washington, D.C. -
Weather coverage and studies.

U.S. Navy, Norfolk, Virginia -
Recovery operations.

Arnold Engineering Development Center, Tullahoma, Tennessee -
High-speed wind tunnel tests.

Air Force Systems Command, Space Systems Division, Inglewood, California -
Project manager for Atlas launch vehicle.

Army Missile Command, Huntsville, Alabama -
Project manager for the Redstone launch vehicle.

Aviation Medical Acceleration Laboratory Johnsville, Pennsylvania -
Astronaut centrifuge training and spacecraft couch and restraint harness testing.

El Centro Naval Parachute Test Facility, El Centro, California -
  Parachute drop test program.

Wright Air Development Center, Dayton, Ohio -
  Acceleration tests, noise and vibration study support, astronaut medical examinations, and pressure suit indoctrination.

Aero Medical Field Laboratory, Holloman Air Force Base, New Mexico -
  Furnished and trained primates for the Mercury animal program.

U.S. Air Force Survival School, Stead Air Force Base, Reno, Nevada -
  Astronaut desert survival training.

Aerospace Medical Division, San Antonio, Texas -
  Cooperation in bioscience experiments in the Little Joe flights.

China Lake Naval Ordnance Test Station, China Lake, California -
  Fairing-adapter sled tests.

Pensacola Naval Air Station, Pensacola, Florida -
  Astronaut egress training and swimmer training in support of Mercury spacecraft recovery operations, and flotation collar fabrication.

Elgin Air Force Base, Pensacola, Florida -
  Astronaut survival training.

Air Force Chart and Information Center, St. Louis, Missouri -
  Worldwide Mercury tracking network maps.

U.S. Army, Fort Eustis, Newport News, Virginia -
  Supplied LARC vehicle for recovery.

Public Health Service, Washington, D.C. -
  Supplied medical monitor personnel.

U.S. Marine Corps Air Station, Cherry Point, North Carolina -
  Recovery helicopters.

Military Air Transport Service, Scott Air Force Base, Illinois -
  Air transportation.

White Sands Missile Range, White Sands, New Mexico -
  Tracking facilities.

Pacific Missile Range, Point Mugu, California -
  Tracking facilities.

Naval Air Station, Corpus Christi, Texas -
  Tracking facilities.

Department of State, Washington, D.C. -
  Government-to-government negotiations for overseas sites in Mercury worldwide tracking network.

Navy Daingerfield Test Facility, Daingerfield, California -
  Spacecraft afterbody shingle heat resistance and dynamic pressure test.

Navy Aircrew Equipment Laboratory, Philadelphia, Pennsylvania -
  Reentry-heat-pulse orientation for the astronauts and environmental systems training.

U.S. Air Force Flight Test Center, Edwards Air Force Base, California -
  Assistance to NASA Flight Research Center in parachute drop tests and reefing parameters.

Army Audit Office, Pasadena, California -
  Audit services.
District Coast Guard, San Francisco, California -
  Support services for project.
Headquarters, U.S. Air Force, Washington, D.C. -
  Airlift services and data reduction.
Walter Reed Army Medical Center, Washington, D.C. -
  Animal test program.
U.S. Navy Comptroller, Washington, D.C. -
  Materiel inspection services.
U.S. Navy Bureau of Ships, Washington, D.C. -
  Command receiver equipment and installation, planned recovery, contingency recovery, and modification of destroyer davits for spacecraft recovery.
U.S. Navy Bureau of Weapons, Washington, D.C. -
  Consultation, planned recovery, and recovery operations.
U.S. Navy Research Laboratory, Washington, D.C. -
  Consultive services.
U.S. Navy Weapons Plant, Washington, D.C. -
  Packaging of hardware.
U.S. Marine Corps Air Facility, Jacksonville, Florida -
  Pilot test facility.
Air Rescue Service, Military Air Transport Service, Orlando, Florida -
  Recovery operation support.
U.S. Navy Bureau of Weapons, Pensacola, Florida -
  Fabrication of auxiliary flotation collar for Mercury spacecraft.
U.S. Navy School of Aviation Medicine, Pensacola, Florida -
  Radiation monitoring for Mercury flights.
Naval Air Station, Pensacola, Florida -
  Spacecraft test flotation collar.
U.S. European Command, Paris, France -
  Contingency recovery.
U.S. Army, Europe, Heidelberg, Germany and Weisbaden, Germany -
  Contingency recovery.
U.S. Air Forces, Europe, Weisbaden, Germany -
  Contingency recovery.
U.S. Air Force Pacific Air Forces, Hickam Air Force Base, Hawaii -
  Contingency recovery and photographic services.
U.S. Air Force Communications Service, Scott Air Force Base, Illinois -
  Communication services during recovery.
U.S. Navy Bureau of Weapons, St. Louis, Missouri -
  Issued Government bill of lading for Contract NAS 5-59, Mercury spacecraft, also contracting support, etc.
U.S. Army Audit Agency, New York, New York -

Audit services.

U.S. Navy Laboratory, Dalgren, Virginia -
  Cartridges for Mercury antenna.

Air Weather Service, Scott Air Force Base, Illinois -
  Weather surveillance flights preceding Mercury manned orbital mission.

Tactical Air Command, Langley Air Force Base, Virginia -
  C-130 aircraft for spacecraft test drops and photographic coverage.

U.S. Army Research and Engineering Laboratory, Natick, Massachusetts -
  Preliminary studies of calorie and water requirements for astronauts during orbital flight.

U.S. Air Force Surgeon General's Office, Washington, D.C. -
  Compilation of medical monitors' training program.